# THE BIG BOOK OF UFO FACTS, FIGURES & TRUTH

THE

# BIG BOOK OF

## UFO FACTS, FIGURES & TRUTH

A Comprehensive Examination

STEPHEN SPIGNESI &
WILLIAM BIRNES

Skyhorse Publishing

Copyright © 2019 by Stephen Spignesi & William Birnes

All rights reserved. No part of this book may be reproduced in any manner without the express written consent of the publisher, except in the case of brief excerpts in critical reviews or articles. All inquiries should be addressed to Skyhorse Publishing, 307 West 36th Street, 11th Floor, New York, NY 10018.

Skyhorse Publishing books may be purchased in bulk at special discounts for sales promotion, corporate gifts, fund-raising, or educational purposes. Special editions can also be created to specifications. For details, contact the Special Sales Department, Skyhorse Publishing, 307 West 36th Street, 11th Floor, New York, NY 10018 or info@skyhorsepublishing.com.

Skyhorse® and Skyhorse Publishing® are registered trademarks of Skyhorse Publishing, Inc.®, a Delaware corporation.

Visit our website at www.skyhorsepublishing.com.

10 9 8 7 6 5 4 3

Library of Congress Cataloging-in-Publication Data is available on file.

Cover design by Rain Saukas

Print ISBN: 978-1-51072-085-5
Ebook ISBN: 978-1-51072-086-2

Printed in United States of America

# DEDICATION

*Stephen Spignesi*
This book is dedicated to my friends
Colin Andrews & Synthia Andrews

*William Birnes*
I dedicate this book to my wife and co-host
on *Future Theater*, Nancy Birnes,

and to my wonderful cast and crew on the History Channel's
*UFO Hunters* series.

# CONTENTS



## PART I
## THE PROJECT BLUE BOOK MYSTERIES



## PART II
## IN PLAIN SIGHT: UFOS IN THE SKY FROM ANCIENT TO MODERN TIMES

## PART III
## A UFO WHO'S WHO



## PART IV
## ALIEN ANCESTRY

**PART V**

## TAKEN: ABDUCTIONS BY ALIENS



**PART VI**

## ROSWELL: THE UFO MYSTERY THAT WON'T GO AWAY



**PART VII**

## UFOS AND THE BIBLE

## PART VIII
## AS WEIRD AS IT GETS



## PART IX
## POPULAR CULTURE AND THE UFO PHENOMENON

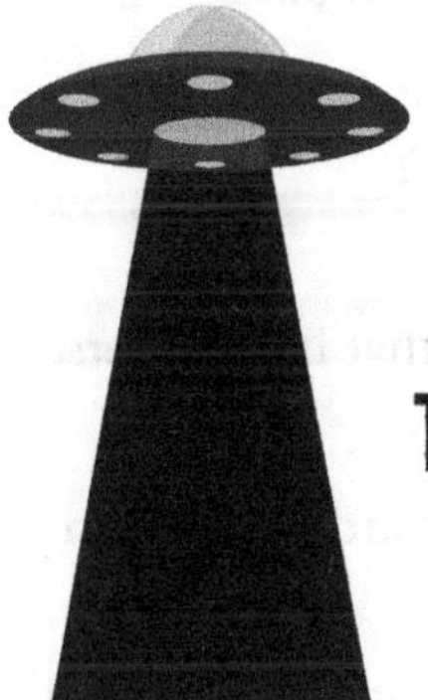

# INTRODUCTION

## This Used to Be Our Playground

To introduce this new edition of this book, we once again would like to briefly step aside and defer to two men who share the name of John.

The first is author John White, from his book, *The Meeting of Science and Spirit:*

> *If UFOs do not exist, then neither do stars. Some of the same types of evidence which long ago convinced us stars exist suggest that UFOs are a reality as well: visual sightings around the world by competent witnesses, photographs and films of their appearance in the sky (pronounced genuine by many reputable experts), and observable secondary effects such as skin burns (equivalent to sunburn) and other radiational effects seen on those who claim to have had UFO encounters.*
>
> *The evidence is in: UFOs are real.*

The second is seventeenth-century poet John Donne, from his essay, "Meditations Upon Our Human Condition":

> *Men that inhere upon nature only, are so far from thinking, that there is anything singular in this world, as that they will scarce think, that this world itself is singular, but that every planet, and every star, is another world like this; they find reason to*

> *conceive, not only a plurality in every species in the world, but a plurality of worlds; so that the abhorrers of solitude, are not solitary; for God, and nature and reason concur against it.*

It is incredibly arrogant on the part of mankind to believe that in a universe unimaginably vast, humans are the only life form in existence.

Perhaps the chapters that follow will mitigate this foolhardy—and ultimately dangerous—terran narcissism.

The galaxy—and beyond—is not only mankind's playground anymore . . . and the skies await us.

Godspeed.

STEPHEN SPIGNESI

WILLIAM BIRNES

# THE BIG BOOK OF UFO FACTS, FIGURES & TRUTH

# PART I

# THE PROJECT BLUE BOOK MYSTERIES

# Currently Uninvolved: NASA's Official Statement on the US Government's Involvement in Investigating UFO Reports

This is what the US government has to say about its current involvement in the UFO controversy. Your reaction to this statement will depend in a large part on your personal point of view regarding the UFO phenomenon.

## FS-1997-01-017-HQ: The US Government and Unidentified Flying Objects

No branch of the United States Government is currently involved with or responsible for investigations into the possibility of alien life on other planets or for investigating Unidentified Flying Objects (UFOs). The US Air Force (USAF) and the National Aeronautics and Space Administration (NASA) have had intermittent, independent investigations of the possibility of alien life on other planets; however, none of these has produced factual evidence that life exists on other planets, nor that UFOs are related to aliens. From 1947 to 1969, the Air Force investigated UFOs; then in 1977, NASA was asked to examine the possibility of resuming UFO investigations. After studying all of the facts available, it was determined that nothing would be gained by further investigation, since there was an absence of tangible evidence. In October 1992, NASA was directed by Congress to begin a detailed search for artificial radio signals from other civilizations under the NASA Towards Other Planetary Systems (TOPS)/High Resolution Microwave

Survey (HRMS) program (also known as the Search for Extraterrestrial Intelligence project). Congress directed NASA to end this project in October 1993, citing pressures on the US Federal budget. The HRMS did not detect any confirmed signal before it was stopped. However, similar work will continue in a more limited manner through efforts of private groups and through academic institutions. The Search for Extraterrestrial Intelligence Institute (SETI Institute) in Mountain View, CA, effectively replaced the Government project, borrowing the signal processing system from NASA. The SETI Institute is a nonprofit corporation conducting research in a number of fields including all science and technology aspects of astronomy and planetary sciences, chemical evolution, the origin of life, biological evolution, and cultural evolution.

During several space missions, NASA astronauts have reported phenomena not immediately explainable; however, in every instance NASA determined that the observations could not be termed "abnormal" in the space environment. The 1947 to 1969 USAF investigations studied UFOs under Project Blue Book. The project, headquartered at Wright-Patterson Air Force Base, Ohio, was terminated December 17, 1969. Of the total of 12,618 sightings reported to Project Blue Book, 701 remain "unidentified."

The decision to discontinue UFO investigations was based on an evaluation of a report prepared by the University of Colorado entitled, "Scientific Study of Unidentified Flying Objects;" a review of the University of Colorado's report by the National Academy of Sciences; previous UFO studies; and Air Force experience investigating UFO reports during the 1940s, '50s, and '60s. As a result of experience, investigations, and studies since 1948, the conclusions of Project Blue Book were: (1) no UFO reported, investigated, and evaluated by the Air Force was ever a threat to our national security; (2) there was no evidence submitted to, or discovered by, the Air Force that sightings categorized as "unidentified" represented technological developments or principles beyond the range of modern scientific knowledge; and (3) there was no evidence indicating that sightings categorized as "unidentified" were extraterrestrial vehicles.

With the termination of Project Blue Book, the USAF regulation establishing and controlling the program for investigating and analyzing UFOs was rescinded. Documentation regarding the former Project Blue Book investigation was permanently transferred to the Modern Military Branch, National Archives and Records Service, in Washington, DC 20408, and is available for public review and analysis.

Since the termination of Project Blue Book, nothing has occurred that would support a resumption of UFO investigations by the USAF or NASA. Given the current environment of steadily decreasing defense and space budgets, it is unlikely that the Air Force or NASA will become involved in this type of costly project in the foreseeable future.

Since neither NASA nor the Air Force is engaged in day-to-day UFO research, neither one reviews UFO-related articles intended for publication, evaluates UFO-type spacecraft drawings, or accepts accounts of UFO sightings or applications for employment in the field of aerial phenomena investigation.

In fact, when Dean Condon of the University of Colorado wrote his introduction to the University of Colorado Study of UFOs in which he discounted them as anything but alien space ships, he was following the instructions of the United States Air Force, who wanted to get out of the UFO business by shutting down Project Blue Book. But the real secret behind this conspiracy to cover up the USAF continuing investigation into the UFO phenomena was a function of an investigation undertaken by attorney Roy Cohn, who was working for Senator Joe McCarthy on the Senate hearings into Communist activities. Professor Condon was a physics professor who was teaching quantum theory, which was considered "subversive" by the McCarthy cadre of supporters. When Cohn accused Condon of being a subversive because of his scholarship into quantum physics, Condon, in order to get back into the good graces of the US government, agreed to disparage the concept of UFOs so as to do the government's bidding and shut down the Air Force investigation into the phenomenon.

Even though Blue Book was ostensibly shut down, the US military continued to collect reports of UFO sightings, some of which they investigated.

A case in point was the sighting by two F-18 Hornet Navy pilots from the aircraft carrier *USS Nimitz* on a routine mission off San Diego. Their sighting, captured on gun camera video, was a startling observation of an unknown craft displaying extraordinary aeronautics. However, there was no military response, and the Pacific command seemed unfazed by the sighting. Did this mean that the Pentagon knew exactly what the pilots had observed and found no threat there?

## UFO Points of Contact:

- For further information on the Search for Extraterrestrial Intelligence, please contact the SETI Institute, 2035 Landings Drive, Mountain View, CA 94043, (415) 960-4530.

- News media requiring Project Blue Book files should contact the National Archives Public Affairs Office, (202) 501-5525. Public queries should be addressed to the Project Blue Book archivist at (202) 501-5385. For queries not related to Project Blue Book, contact the National Archives receptionist at (202) 501-5400. Documentation is available from: Modern Military Branch, National Archives and Records Service, Eighth Street and Pennsylvania Avenue, NW, Washington, DC 20408.

- The Air Force publication, "The Roswell Report: Fact vs. Fiction in the New Mexico Desert," a lengthy document providing all of the details available from the Air Force on the Roswell incident, is available for $52 from the US Government Printing Office, Superintendent of Documents, Mail Stop: SSOP, Washington, DC 20402-9328.

- There are a number of universities and professional scientific organizations that have considered UFO phenomena during periodic meetings and seminars. A list of private organizations interested in aerial phenomena may be found in Gale's Encyclopedia of Associations.

- Persons wishing to report UFO sightings are advised to contact law enforcement agencies.

# 20 Years of Project Blue Book Mysteries 218 Sightings from 1947–1967 by Military Personnel

*The phenomenon reported is something real and not visionary or fictitious. It is recommended that . . . Headquarters, Army Air Forces issue a directive assigning a priority, security classification and Code Name for a detailed study of this matter.*

—Lieutenant General Nathan F. Twining
head of the Air Materiel Command, in a September 23, 1947 letter to the commanding general of the Army Air Forces regarding the initiation of Project Sign

These are the "Project Blue Book" UFO sightings that even the United States government cannot explain. And what is most disconcerting for the Feds is that all of these 218 unexplained sightings were made by—and properly reported by—*trained military personnel, including many, many pilots of all ranks and armed services,* from the lowliest grounds crewman to Major-Generals.

After reading eyewitness account after eyewitness account of objects (including disks, ovals, and other odd-shaped craft) zig-zagging at impossible speeds, flying around military aircraft and performing incredible maneuvers, all apparently under intelligent control, one cannot help but be flabbergasted by the significance of these reports.

As indicated, adding enormous credibility to these sightings is the fact that they were all made by trained military personnel. Sober-minded, serious, alert, and *observant* men—professional men the United States government spent millions training and educating so they could make precisely the kind of accurate assessments and reports they made in these cases. (Also included are sightings by a couple of US politicians and other government personnel—a US District Court reporter, a Forest Ranger, etc.—whose credibility would normally be considered unimpeachable.)

These sightings are in chronological order and were compiled from the almost six hundred unexplained Project Blue Book UFO sightings remaining when the Project shut down in 1967.

What follows is a brief history of the three most important, official UFO-related United States government projects: Project Sign, Project Grudge, and Project Blue Book.

On December 23, 1947, Project Sign was established to collect, collate, evaluate, and disseminate all information pertaining to UFO sightings and other unexplained aerial phenomena. In February of 1948, Project Sign completed operations and issued a statement that read, "No definite and conclusive evidence is yet available that would prove or disprove the existence of these UFOs as real aircraft of unknown and unconventional configuration."

In December 1948, Project Sign was changed to Project Grudge, which, according to the project's last director, Edward Ruppelt, evaluated reports "on the premise that UFOs couldn't exist. No matter what you see or hear, don't believe it." Project Grudge ultimately reviewed 244 sightings, of which 23 percent were unexplainable, but the project lapsed into inactivity and was shut down on December 27, 1949.

In March 1952, Project Blue Book was begun and continued investigating UFO reports until 1967.

Project Blue Book's stated goals were as follows: to find an explanation for all the reported sighting[s] of UFOs, to determine if the UFO posed any security threat to the United States, and to determine if UFOs exhibit[ed] any advanced technology which the US could utilize.

Project Blue Book investigators chronicled and investigated over twelve thousand UFO reports and explained ("explained away," a great many suggested) over eleven thousand of these sightings. Of the close to six hundred that were officially declared as unexplained, the 218 listed here are the ones made by the military's own.

Some of these sightings are incredible. Check out number 9, for instance. Or numbers 21, 28, 91, 103, 173, 194, 207, or 217.

After you review these and all the other sightings listed here, make up your own mind as to how you might "explain away" these unexplained reports.

## Sighting Key

- **a.** Date and Time of Sighting
- **b.** Location of Sighting
- **c.** Witnesses
- **d.** Duration of Sighting
- **e.** Description of Sighting

## 1947

**1.** a. Sunday, July 6, daytime.
b. Fairfield-Suisan Air Base in California.
c. Army Air Forces Captain and Mrs. James Burniston.
d. 1 minute.
e. One object having no wings or tail rolled from side-to-side three times and then flew away very fast to the southeast.

**2.** a. Tuesday, July 8 at 9:30 a.m.
b. Muroc Air Base in California.
c. First Lieutenant Joseph McHenry, Technical Sergeant Ruvolo, Staff Sergeant Nauman, and Miss Janette Scotte.
d. Brief.
e. Two silver and apparently metallic disk-shaped or spherical objects flew a wide circular pattern; one of them later flew a tighter circle.

**3.** a. Wednesday, July 9 at 12:17 p.m.
b. Meridian, Idaho.
c. Idaho statesman aviation editor and former Army Air Force B-29 pilot Dave Johnson.

d. More than 10 seconds.

e. A black disk, which stood out against the clouds and which was seen from an Idaho Air National Guard AT-6, made a half-roll and then a stair-step climb.

**4.** a. Tuesday, July 29 at 2:50 p.m.

b. Hamilton Air Base, California.

c. Assistant Base Operations Officer Captain William Rhyerd, ex-Army Air Force B-29 pilot Ward Stewart.

d. Unknown.

e. Two round, shiny, white objects with estimated 15 to 25 foot diameters, flew three to four times the apparent speed of a P-80, also in view. One object flew straight and level; the other weaved from side-to-side like an escort fighter.

**5.** a. Tuesday, October 14 at noon.

b. 11 miles north-northeast of Cave Creek, Arizona.

c. Ex-Army Air Force fighter pilot J. L. Clark, civilian pilot Anderson, and a third man.

d. 45–60 seconds.

e. One 3-foot "flying wing"-shaped object, which looked black against the white clouds and red against the blue sky, flew straight at an estimated 380 m.p.h., at 8–10,000 feet, from the northwest to the southeast.

## 1948

**6.** a. Thursday, September 23 at noon.

b. San Pablo, California.

c. Retired US Army Colonel Horace Eakins and Sylvester Bentham.

d. Unknown.

e. Two objects were seen. One was a buff or grey rectangle with vertical lines; the other was a translucent "amoeba" with a dark spot near the center. The arms of the "amoeba" undulated and both objects traveled very fast.

7. a. Friday, December 3 at 8:15 p.m.

b. Fairfield-Suisan Air Force Base, California.

c. US Air Force Sergeant, control tower operator.

d. 25 seconds.

e. One round, white light flew with varying speed and a bouncing motion, and finally made a rapid, erratic climb.

## 1949

8. a. Tuesday, January 4 at 2:00 p.m.

b. Hickam Field, Hawaii.

c. US Air Force pilot Captain Paul Storey, on the ground.

d. Brief.

e. One flat white, elliptical object with a matte top circled while oscillating to the right and left, and then sped away.

9. a. Thursday, January 27 at 10:20 p.m.

b. Cortez-Bradenton, Florida.

c. Acting chief of the Aircraft Branch, Elgin Air Force Base, Captain Sames, and Mrs. Sames.

d. 25 minutes.

e. A cigar-shaped object as long as two Pullman cars and with seven lighted square windows, threw sparks, descended, and then climbed with a bouncing motion at an estimated 400 m.p.h.

**10.** a. Thursday, March 17 at 7:52 p.m.
b. Camp Hood, Texas.
c. Guards of the Second Armored Division.
d. 1 hour.
e. Eight large green, red and white flare-like objects flew in generally straight lines for the entire duration of the sighting.

**11.** a. Monday, April 4 at 10:20 p.m.
b. Merced, California.
c. William Parrott, former Air Force pilot and Major.
d. 35 seconds.
e. One generally round object with a curved bottom and dull coloring flew overhead and gave off a clicking sound.

**12.** a. Thursday, May 5 at 11:40 a.m.
b. Ft. Bliss, Texas.
c. Army officers Major Day, Major Olhausen, Captain Vaughn.
d. 30–35 seconds.
e. Two oblong white disks, flying at an estimated 200–250 m.p.h., made a shallow turn during the observation.

**13.** a. Monday, May 9 at 2:30 p.m.
b. Tucson, Arizona.
c. Master Sergeant Troy Putnam.
d. Brief.
e. Two round, flat silvery objects, estimated to be 25 feet in diameter, flew 750–1,000 m.p.h. in a banked but steady manner.

## 1950

**14.** a. Sunday, February 5 at 5:10 p.m.

b. Teaticket, Massachusetts.

c. Former US Navy fighter pilot Marvin Odom, US Air Force Lieutenant Philip Foushee, a pilot from Otis Air Force Base, and two others.

d. 5 minutes.

e. Two thin, illuminated cylinders, one of which dropped a fireball, maneuvered together and then disappeared high and fast.

**15.** a. Monday, March 27, 1980 at 10:30 a.m.

b. Motobo, Okinawa.

c. US Air Force radar operator Corporal Bolfango.

d. 2 minutes.

e. An unknown object was tracked on radar while it was stationary, and then also as it moved away at a speed of approximately 500 m.p.h.

**16.** a. Tuesday, March 28 at 3:15 p.m.

b. Santiago, Chile.

c. Master Sergeant Patterson of the office of the US Air Attaché.

d. 5–10 seconds.

e. One white object was observed through binoculars while it flew high and fast, crossing 30 degrees of sky.

**17.** a. Friday, April 14 at 2:30 p.m.

b. Ft. Monmouth, New Jersey.

c. Army Master Sergeant James.

d. 3–4 minutes.

e. Four rectangular, amber objects, about 3 feet by 4 feet in size, changed speed and direction rapidly. The group of objects rose and fell during the sighting.

**18.** a. Sunday, August 20 at 1:30 p.m.
b. Nicosia, Cyprus.
c. US Air Force MATS liaison officer Lieutenant William Ghormley, Colonel W. V. Brown, and Lieutenant Col. L. W. Brauer.
d. 15–20 seconds.
e. One small, round, bright object flew fast, straight, and level for the duration of the sighting.

**19.** a. Friday, August 25 at 8:00 p.m.
b. Approximately 250 miles southwest of Bermuda.
c. B-29 radarman Staff Sergeant William Shaffer.
d. 20 minutes.
e. A B-29 followed an unidentified target, then passed it at a one-quarter mile distance. There was radar observation, plus a possible blue streak 3 minutes later. The target followed the plane for 5 minutes, then passed it and sped away.

**20.** a. Thursday, September 21 at 9:52 a.m.
b. Provincetown, Massachusetts.
c. MIT research associate and Air National Guard Major M. H. Ligda.
d. 1 minute.
e. There was a confirmed radar tracking of one object during MIT tracking of a US Air Force flight of F-84 or F-86 jet fighters. The object's speed was approximately 22 miles per minute (1,300 m.p.h.). The object made a turn of 11–12 gravities acceleration during the observation.

**21.** a. Monday, October 23 at 12:42 p.m.
b. Bonlee, North Carolina.
c. Ex-US Air Force pilot Frank Risher.
d. 40 seconds.
e. One aluminum object shaped like a dirigible or a Convair C-99 cargo plane, with three portholes, arrived from the southeast, hovered for a few seconds and flew away to the south-southeast at the end of the sighting.

## 1951

**22.** a. Friday, January 12 at 10:00 p.m.
b. Ft. Benning, Georgia.
c. US Army Second Lieutenant A.C. Hale.
d. 20 minutes.
e. One light with a fan-shaped wake remained motionless like a star and then sped away.

**23.** a. Thursday, February 1 at 5:10 p.m.
b. Johnson Air Base, Japan.
c. The pilot and radar operator of an F-82 night fighter.
d. Brief.
e. One amber light made three or four 360 degree turns to the right, reversed toward the F-82, and then climbed out of sight.

**24.** a. Sunday, February 25 or Monday, February 26 at 7:10 a.m.
b. Ladd Air Force Base, Alaska.
c. US Air Force Sergeant J. B. Sells.
d. 60–90 seconds.
e. One dull gray, metallic object, estimated to be 120 feet long and 10–12 inches thick, hovered, puffed smoke and sped away.

**25.** a. Saturday, March 10 at 9:51 a.m.
b. Chinnampo, Korea.
c. The crew of a US Air Force B-29 bomber, including scanners and a tail gunner.
d. Brief.
e. A large red-yellow glow burst and became blue-white.

**26.** a. Tuesday, March 13 at 3:20 p.m.
b. McClellan Air Force Base, California.
c. US Air Force First Lieutenant B. J. Hastie, Mrs. Rafferty.
d. 2 minutes.
e. A cylinder with twin tails, 200 feet long and 90 feet wide, turned north and flew away at incredible speed.

**27.** a. Friday, June 1 at 4:20 a.m.
b. Niagara Falls, New York.
c. Master Sergeant H. E. Sweeney, and two enlisted men.
d. 30–40 seconds.
e. One glowing yellow-orange, saucer-shaped object with arc-shaped wings, flew straight up.

**28.** a. Tuesday, July 24 at 7:10 a.m.[?].
b. Portsmouth, New Hampshire.
c. Hanscom Air Force Base Operations Officer Captain Cobb, Corporal Fein.
d. 20 seconds.
e. One 100–200 foot tubular object, five times long as it was wide, with fins at one end, and colored greyish with many black spots flew 800–1,000 m.p.h. at 1–2,000 feet altitude, leaving a faint swath.

**29.** a. Thursday, September 6 at approximately 7:20 p.m.
b. Claremont, California.

c. Staff Sergeant W. T. Smith, Master Sergeant L. L. Duel.
d. 3–4 minutes.
e. Six orange lights in an irregular formation flew straight and level into a coastal fog bank.

**30.** a. Friday, September 14 at 9:30 p.m.
b. Goose Bay, Labrador, Canada.
c. Technical Sergeant W. B. Maupin, Corporal J. W. Green.
d. Over 15 minutes.
e. Three objects were tracked on radar. Two were on a collision course, then one evaded to the right upon the request, by radio, of one of the radar operators. No aircraft were known to be in the area. A third unidentified object tracked, then joined the first two.

**31.** a. Wednesday, October 3 at 10:27 p.m.
b. Kadena, Okinawa.
c. Radar operators Sergeant M. W. Watson and Private Gonzales, and one other unidentified Sergeant.
d. Brief.
e. One large, sausage-shaped blip was tracked at an estimated 4,800 m.p.h.

**32.** a. Saturday, November 24 at 3:53 p.m.
b. Mankato, Minnesota.
c. US Air Force or Air National Guard pilots W. H. Fairbrother and D. E. Stewart in P-51 Mustangs.
d. 5 seconds.
e. One milky white object shaped like a Northrop flying wing—it had a broad, slightly swept-back wing with no fuselage or tail and an estimated 8-foot span—flew straight and level for the duration of the sighting.

## 1952

**33.** a. Monday, February 11 at 3:00 a.m.
b. Pittsburgh, Pennsylvania.
c. Captain G. P. Arns and Major R. J. Gedson flying a Beech AT-ll trainer.
d. 1 minute.
e. One yellow-orange comet-shaped object flew a straight and level flight and pulsed flame for 1–2 seconds.

**34.** a. Saturday, February 23 at 11:15 p.m.
b. Over North Korea.
c. The Captain and navigator of a B-29.
d. 45 seconds.
e. One bluish cylinder with a tail and rapid pulsations, three times long as it was wide, came in high and fast, made several turns, and then leveled out under a B-29 which was evading mild anti-aircraft fire.

**35.** a. Thursday, March 20 at 10:42 p.m.
b. Centreville, Maryland.
c. World Wars I and II veteran A. D. Hutchinson and his son.
d. 30 seconds.
e. One dull orange-yellow saucer-shaped light flew straight and level and very fast the duration of the sighting.

**36.** a. Sunday, March 23 at 6:56 and 7:00 p.m.
b. Yakima, Washington.
c. The pilot and radar operator of an F-94 jet interceptor.
d. 45 seconds.
e. On either occasion, a stationary red fireball increased in brightness and then faded over the duration of the sighting. Project

Blue Book Status Report #7 (May 31, 1952) says the target was also tracked by ground radar at 78 knots (90 m.p.h.) and alternately at 22,500 feet and 25,000 feet altitude.

**37.** a. Monday, March 24 at 8:45 a.m.
b. 60 miles west of Pt. Concepcion, California.
c. A B-29 navigator and radar operator.
d. 20–30 seconds.
e. One target was tracked at an estimated 3,000 m.p.h.

**38.** a. Saturday, March 29 at 11:20 a.m.
b. 20 miles north of Misawa Air Force Base, Japan.
c. US Air Force pilot Brigham, flying an AT-6 trainer.
d. 10 seconds.
e. One small, very thin, shiny metallic disk flew alongside the AT-6, then made a pass at an F-84 jet fighter, flipped on its edge, fluttered 20 feet from the F-84's fuselage and then flipped in the slipstream.

**39.** a. Friday, April 4 at 7:30 p.m.
b. Duncanville, Texas.
c. Two radar operators of the 147th AC&W Squadron.
d. 1 minute.
e. One object was tracked by radar at an estimated 2,160 m.p.h.

**40.** a. Saturday, April 12 at 9:30 p.m.
b. North Bay, Ontario, Canada.
c. Royal Canadian Air Force Warrant Officer E. H. Rossell, Flight Sergeant R. McRae.
d. 2 minutes.
e. One round amber object flew fast, stopped, reversed direction, and climbed away at a 30 degree angle during the duration of the sighting.

**41.** a. Monday, April 14 at 6:34 p.m.
b. Memphis, Tennessee.
c. US Navy pilots Lieutenant Junior Grade Blacky, Lieutenant Junior Grade O'Neil.
d. 45–60 seconds.
e. One inverted bowl, 3 feet long and 1 foot high, with vertical slots, flew fast, straight and level, 100 yards from the observers' aircraft for the duration of the sighting.

**42.** a. Thursday, April 17 at 3:05 p.m.
b. Yuma, Arizona.
c. A group of Army weather observation students, including several graduate engineers.
d. 7 seconds.
e. One flat-white, circular object flew with an irregular trajectory and a brief trail.

**43.** a. Friday, April 18, time unknown.
b. Yuma, Arizona.
c. Two Army weather observation students.
d. 5–10 seconds.
e. One flat-white, circular object flew in a very erratic manner.

**44.** a. Tuesday, April 22 at 9:00 p.m.
b. Naha Air Force Base, Okinawa.
c. The crew of a B-29 bomber on the ground.
d. 10 minutes.
e. Five elliptical objects, each with a white light that blinked every 1–2 seconds, performed erratic maneuvers for the duration of the sighting.

**45.** a. Thursday, April 24 at 5:00 a.m.
b. Bellevue Hill, Vermont.
c. The crew of a US Air Force C-124 transport plane.
d. 3–4 minutes.
e. Three circular, bluish objects in loose "fingertip" formation twice flew parallel to the C-124 during the duration of the sighting.

**46.** a. Thursday, April 24 at 8:10 p.m.
b. Clovis, New Mexico.
c. US Air Force Light Surgeon Major E. L. Ellis.
d. 5 minutes.
e. Many orange-amber lights, sometimes separate, sometimes fused, behaved erratically with their speed varying from motionless to very fast during the sighting.

**47.** a. Sunday, April 27 at 8:30 p.m.
b. Yuma, Arizona.
c. Off-duty control tower operator Master Sergeant G. S. Porter and Mrs. Porter.
d. 2 hours.
e. There were seven sightings of one bright red disk, and one sighting of two disks flying in formation during the duration of the sighting. All the disks appeared as large as fighter planes and were all seen below an 11,000 foot overcast.

**48.** a. Tuesday, April 29 at 1:00 p.m.
b. Goodland, Kansas.
c. B-29 bombardier Lieutenant R. H. Bauer.
d. 2 seconds.
e. One white fan-shaped light pulsed three to four times per second.

**49.** a. Thursday, May 1 at 10:50 a.m.
b. George Air Force Base, California.
c. Three military personnel on the arms range, plus one Lieutenant Colonel 4 miles away.
d. 15–30 seconds.
e. Five flat-white disks about the diameter of a C-47's wingspan (95 feet) flew fast, made a 90 degree turn in a formation of three in front and two behind, and darted around.

**50.** a. Wednesday, May 7 at 12:15 p.m.
b. Keesler Air Force Base, Mississippi.
c. Captain Morris, a Master Sergeant, a Staff Sergeant, and an Airman First Class.
d. 5–10 minutes.
e. An aluminum or silver cylindrical object was seen to dart in and out of the clouds ten times during the duration of the sighting.

**51.** a. Wednesday, May 14 at 7:00 p.m.
b. Mayaquez, Puerto Rico.
c. Ex-US Air Force pilot and attorney Mr. Stipes, Sr. Garcia-Mendez.
d. 30 minutes.
e. Two shining orange spheres were seen: one remained stationary while the other darted away and back for the duration of the sighting.

**52.** a. Tuesday, May 20 at 10:10 p.m.
b. Houston, Texas.
c. US Air Force pilots Captain J. Spurgin and Captain B. B. Stephan.
d. 90 seconds.

e. One bright or white oval object moved from side-to-side while making a gradual turn.

**53.** a. Thursday, May 29 at 7:00 p.m.
b. San Antonio, Texas.
c. US Air Force pilot Major D. W. Feuerstein on the ground.
d. 14 minutes.
e. One bright tubular object tilted from horizontal to vertical for 8 minutes, then over the next 6 minutes slowly returned to horizontal, again tilted vertical, accelerated, appeared to lengthen and then turned red.

**54.** a. Sunday, June 1 at 1:00 p.m.
b. Walla Walla, Washington.
c. Ex-military pilot Reserve Major W. C. Vollendorf.
d. 7 seconds.
e. One oval object with a "definite airfoil" performed a fast climb.

**55.** a. Monday, June 2, time unknown.
b. Fulda, West Germany.
c. First Lieutenant John Hendry, a photo-navigator on an RB-26C reconnaissance bomber.
d. Unknown.
e. One porcelain-white object flew very fast.

**56.** a. Thursday, June 5 at 6:45 p.m.
b. Albuquerque, New Mexico.
c. Staff Sergeant T. H. Shorey.
d. 6 seconds.
e. One shiny round object flew five to six times as fast as an F-86 jet fighter.

**57.** a. Thursday, June 5 at 11:00 p.m.
b. Offutt Air Force Base, Omaha, Nebraska.
c. Strategic Air Command top secret control officer and former OSI agent Second Lieutenant W. R. Soper, and two other persons.
d. 4½ minutes.
e. One bright red object remained stationary for the duration of the sighting before speeding away with a short tail.

**58.** a. Saturday, June 7 at 11:18 a.m.
b. Albuquerque, New Mexico.
c. The crew of B-25 bomber #8840 flying at 11,500 feet altitude.
d. Brief.
e. One rectangular aluminum object, measuring approximately 6 feet by 4 feet flew 250–300 feet below the B-25.

**59.** a. Thursday, June 12 at 7:30 p.m.
b. Fort Smith, Arkansas.
c. A US Army Major and a Lieutenant Colonel, using binoculars.
d. Brief.
e. One orange ball with a tail flew overhead with a low angular velocity.

**60.** a. Thursday, June 12 at 11:26 a.m.
b. Marrakech, Morocco.
c. Technical Sergeant H. D. Adams, operating an SCR-584 radar set.
d. Brief.
e. One unidentified blip tracked at 650 knots (750 m.p.h.) at greater than 60,000 feet altitude.

**61.** a. Friday, June 13 at 8:45 p.m.
b. Middletown, Pennsylvania.
c. Olmstead Air Force Base employee and former control tower operator R. S. Thomas.
d. 2 seconds.
e. One round, orange object traveled south, stopped for 1 second, turned east, stopped for 1 second, and then went down.

**62.** a. Sunday, June 15 at 11:50 p.m.
b. Louisville, Kentucky.
c. Ex-US Navy radar technician Edward Duke.
d. 15 minutes.
e. One large, cigar-shaped object with a blunt front, lit sides, and a red stern, maneuvered in a leisurely fashion for the duration of the sighting.

**63.** a. Monday, June 16 at 8:30 p.m.
b. Walker Air Force Base, New Mexico.
c. US Air Force maintenance specialist Staff Sergeant Sparks.
d. 1 minute.
e. Five or six greyish disks flew in a half-moon formation at 500–600 m.p.h. for the duration of the sighting.

**64.** a. Tuesday, June 17, sometime between 7:30 and 10:20 p.m.
b. McChord Air Force Base, Washington.
c. Many varied military personnel.
d. 15 minutes.
e. From one to five large silver-yellow objects flew erratically, stopping and starting for the duration of the sighting.

**65.** a. Tuesday, June 17 at 1:28 a.m.
b. Cape Cod, Massachusetts.

c. The pilot of a US Air Force F-94 jet interceptor.
d. 15 seconds.
e. A light like a bright star was observed crossing the nose of the airplane.

**66.** a. Thursday, June 19 at 2:37 a.m.
b. Goose Bay, Labrador, Canada.
c. Second Lieutenant A'Gostino and an unidentified radar operator.
d. 1 minute.
e. One red light turned white while wobbling; radar tracked a stationary target during the sighting.

**67.** a. Thursday, June 19 at 2:00 p.m.
b. Yuma, Arizona.
c. US Air Force pilot John Lane.
d. 10 seconds.
e. One round, white object flew straight and level for the duration of the sighting.

**68.** a. Friday, June 20 at 3:03 p.m.
b. Central Korea.
c. Four Marine Corps Captains and pilots of F4U-4B Corsair fighter planes.
d. 60 seconds.
e. One 10–20 foot white or silver oval object made a left-hand orbit at terrific speed during the sighting.

**69.** a. Saturday, June 21 at 12:30 p.m.
b. Kelly Air Force Base, Texas.
c. B-29 bomber flight engineer Technical Sergeant Howard Davis, flying at 8,000 feet altitude.
d. 1 second.

e. One white, flat object with a sharply pointed front, rounded rear, dark blue center and red rim, trailed sparks as it dove past the B-29 at a distance of 500 feet.

**70.** a. Sunday, June 22 at 10:45 p.m.
b. Pyungthek, Korea.
c. Two Marine Corps Sergeants.
d. Brief.
e. One object shooting red flames and having a 4 foot diameter dove at a runway, hovered briefly over a hill, turned 180 degrees, flashed twice and then was gone.

**71.** a. Monday, June 23 at 9:00 p.m.
b. McChord Air Force Base, Washington.
c. Second Lieutenant K. Thompson.
d. 10 minutes.
e. One very large light flew straight and level for the duration of the sighting.

**72.** a. Monday, June 23 at 10:00 a.m.
b. Owensboro, Kentucky.
c. National Guard Lieutenant Colonel O. L. Depp.
d. 5 seconds.
e. Two objects looking like "giant soap bubbles" and reflecting yellow and lavender colors flew in trail for the duration of the sighting.

**73.** a. Monday, June 23 at 6:08 a.m.
b. Location unknown, but information came via Japan Headquarters "CV 4359."
c. A US Air Force pilot of the 18th Fighter-Bomber Group.
d. Brief.

e. One black, coin-shaped object, 15–20 feet in diameter, made an irregular descent.

**74.** a. Thursday, June 26 at 2:45 a.m.
b. Terre Haute, Indiana.
c. US Air Force Second Lieutenant C. W. Povelites.
d. Brief.
e. An undescribed object flew at 600 m.p.h. and then stopped.

**75.** a. Friday, June 27 at 6:50 p.m.
b. Topeka, Kansas.
c. US Air Force pilot Second Lieutenant K. P. Kelly and Mrs. Kelly.
d. 5 minutes.
e. One pulsating red object changed shape from a circular to a vertical oval as it pulsed. It remained stationary for the duration of the sighting before blinking out.

**76.** a. Saturday, June 28 at 4:10 p.m.
b. Nagoya, Japan.
c. US Air Force electronics countermeasures officer Captain T. W. Barger.
d. Brief.
e. One dark blue elliptical-shaped object with a pulsing border flew straight and level at 700–800 m.p.h.

**77.** a. Sunday, June 29 at 5:45 p.m.
b. O'Hare Airport, Chicago, Illinois.
c. Three US Air Force air police officers.
d. 45 minutes.
e. One bright silver, flat oval object surrounded by a blue haze, hovered, then moved very fast to the right and to the left, and then up and down during the duration of the sighting.

**78.** a. Wednesday, July 9 at 12:45 p.m.
b. Colorado Springs, Colorado.
c. US Air Force pilot Major C. K. Griffin.
d. 12 minutes.
e. One luminous white object shaped like an airfoil minus its trailing edge moved slowly and erratically for the duration of the sighting.

**79.** a. Wednesday, July 9 at 3:35 p.m.
b. Rapid City Air Force Base, South Dakota.
c. Staff Sergeant D.P. Foster and three other persons.
d. 5 seconds.
e. Three times during the duration of the sighting a single white, disk-shaped object sped by observers, flying on a straight and level trajectory.

**80.** a. Saturday, July 12 at 9:00 p.m.
b. Kirksville, Missouri.
c. Several radar controllers who were also military officers.
d. Brief.
e. Several large unexplainable blips were tracked on radar moving at 1,500 knots (1,700 m.p.h.). There was no visual sighting.

**81.** a. Wednesday, July 16 at 9:35 a.m.
b. Beverly, Massachusetts.
c. US Coast Guard photographer Shell Alpert.
d. Brief.
e. Four roughly elliptical blobs of light moving in formation were photographed through the window of a photo lab.

**82.** a. Thursday, July 17 at 11:00 a.m.
b. Lockbourne, Ohio.

c. Several Air National Guard employees.
d. 3 hours.
e. One light like a big star was seen for 3 hours, but disappeared when an aircraft approached it. This object was also seen by Air National Guard personnel on the nights of July 20, 22, and 23.

**83.** a. Friday, July 18 at 9:10 p.m.
b. Lockbourne, Ohio.
c. Technical Sergeant Mahone, Airman Third Class Jennings.
d. 1½ minutes.
e. One amber-colored, elliptical-shaped object with a small flame at the rear gave off a resonant beat sound and periodically increased in brightness while flying around very fast during the duration of the sighting.

**84.** a. Friday, July 18 at 9:45 p.m.
b. Patrick Air Force Base, Florida.
c. Three US Air Force officers and four enlisted men.
d. 1 hour.
e. For the duration of the sighting, a series of hovering and maneuvering red-orange lights were observed flying around in a variety of directions.

**85.** a. Saturday, July 19 at 11:35 p.m.
b. Elkins Park, Pennsylvania.
c. US Air Force pilot Captain C. J. Powley and Mrs. Powley.
d. 5–7 minutes.
e. Two star-like lights maneuvered, hovered, and sped around for the duration of the sighting

**86.** a. Monday, July 21 at 6:30 p.m.
b. Wiesbaden, West Germany.

c. US Air Force pilot Captain E. E. Dougher, WAF (Women in the Air Force) Lieutenant J. J. Stong, situated miles apart.
d. 10–15 minutes.
e. Four bright yellowish lights were seen by Dougher to separate, with two climbing and two flying away level in the opposite direction. Stong watched two reddish lights fly in opposite directions.

**87.** a. Monday, July 21 at 10:40 p.m.
b. San Marcos Air Force Base, Texas.
c. One Lieutenant, two Staff Sergeants, and three airmen.
d. 1 minute.
e. One blue circle with a blue trail was seen to hover and then accelerate to near-sonic speed (700+ m.p.h.).

**88.** a. Monday, July 21 at 8:10 p.m.
b. Rockville, Indiana.
c. One military officer and two enlisted men.
d. 3 minutes.
e. One aluminum, delta-shaped object with a vertical fin flew straight and level, and then hovered for the duration of the sighting.

**89.** a. Tuesday, July 22 at 10:47 p.m.
b. Between Boston and Provincetown, Massachusetts.
c. The pilot and radar operator of a US Air Force F-94 jet interceptor.
d. Brief.
e. One round spinning blue light passed the F-94 in flight.

**90.** a. Tuesday, July 22 from 10:50 p.m. to July 23, 12:45 a.m.
b. Trenton, New Jersey.

c. The crews of several US Air Force F-94 jet interceptors from Dover Air Force Base, Delaware.

d. 2 hours.

e. The F-94 crews reported thirteen visual sightings and one radar tracking of blue-white lights during the 2 hours.

**91.** a. Wednesday, July 23 at 8:40 a.m.

b. Pottstown, Pennsylvania.

c. The two-man crews of three US Air Force F-94 jet interceptors.

d. 1–4 minutes.

e. One large silver object, shaped like a long pear with two or three squares beneath it, flew at 150–180 knots (170–210 m.p.h.), while a smaller object, delta-shaped or swept back, flew around it at 1,000–1,500 knots (1,150–1,700 m.p.h.).

**92.** a. Wednesday, July 23 at 12:50 p.m.

b. Altoona, Pennsylvania.

c. The two-man crews of two US Air Force F-94 jet interceptors flying at 35–46,000 feet altitude.

d. 20 minutes.

e. Three cylindrical objects in a vertical stack formation flew at an altitude of 50–80,000 feet for the duration of the sighting.

**93.** a. Wednesday, July 23 at 11:35 p.m.

b. South Bend, Indiana.

c. US Air Force pilot Captain H.W. Kloth.

d. 9 minutes.

e. Two bright blue-white objects flew together, then the rear one veered off at the conclusion of the sighting.

**94.** a. Thursday, July 24 at 3:40 p.m.

b. Carson Sink, Nevada.

c. US Air Force Lieutenant Colonels McGinn and Barton flying in a B-25 bomber.
d. 3–4 seconds.
e. Three silver, delta-shaped objects, each with a ridge along the top, crossed in front of and above the B-25 at high speed.

**95.** a. Saturday, July 26 at 12:15 a.m.
b. Kansas City, Missouri.
c. US Air Force Captain H. A. Stone and the men in the control towers at Fairfax Field and Municipal Airport.
d. 1 hour.
e. One greenish light with red-orange flashes was watched as it descended in the northwest from 40 degrees elevation to 10 degrees elevation.

**96.** a. Saturday, July 26 at 12:05 a.m.
b. Kirtland Air Force Base, New Mexico.
c. Airman First Class J. M. Donaldson.
d. 3–4 seconds.
e. Eight to ten orange balls flew very fast in a V-shaped formation.

**97.** a. Sunday, July 27 from 10:05 a.m. to 10:20 a.m.
b. Selfridge Air Force Base, Michigan.
c. Three B-29 bomber crewmen on the ground.
d. 30 seconds each.
e. Many round, white objects flew straight and level, very fast, beginning with two at 10:05, then one at 10:10, another one at 10:15, and a final one at 10:20.

**98.** a. Monday, July 28 at 10:20 p.m.
b. Heidelberg, West Germany.

c. Sergeant B. C. Grassmoen, Women's Army Corps Private First Class A. P. Turner.
d. 4–5 minutes.
e. One saucer-shaped object giving off shafts of white light and having the appearance of light metal flew slowly, made a 90 degree turn, and then climbed away fast at the end of the sighting.

**99.** a. Monday, July 28 at 6:00 a.m.
b. McGuire Air Force Base, New Jersey.
c. Ground Control Approach radar operator Master Sergeant W. F. Dees, plus several military personnel in the base control tower.
d. 55 minutes.
e. Radar tracked a large cluster of very distinct blips. Visual observation was of oblong objects having neither wings nor tail. The objects made a very fast turn and at one time were in echelon formation.

**100.** a. Monday, July 28 at 2:15 a.m.
b. McChord Air Force Base, Washington.
c. Technical Sergeant Walstead and Staff Sergeant Calkins of the 635th AC&W Squadron.
d. Brief.
e. One dull, glowing, blue-green ball, the size of a dime held at arm's length, flew very fast, straight, and level for the duration of the sighting.

**101.** a. Tuesday, July 29 at 1:30 a.m.
b. Osceola, Wisconsin.
c. The pilot of an F-51 Mustang in flight and radar operators on the ground.
d. 1 hour.

e. Several clusters of up to ten small radar targets and one large target were observed. The small targets moved from southwest to east at 50–60 knots (60–70 m.p.h.), following each other. The large object moved at 600 knots (700 m.p.h.).

**102.** a. Tuesday, July 29 at 2:30 p.m.
b. Langley Air Force Base, Virginia.
c. US Air Force Captain D. G. Moore of the military air traffic control system.
d. 2 minutes.
e. One undescribed object flew toward the air base at an estimated 2,600 m.p.h. below 5,000 feet altitude.

**103.** a. Tuesday, July 29 at 12:35 p.m.
b. Wichita, Kansas.
c. US Air Force shop employees Douglas and Hess at Municipal Airport.
d. 5 minutes.
e. One bright white circular object with a flat bottom flew very fast above the Airport, and then hovered for 10–15 seconds over the Cessna Aircraft Company plant.

**104.** a. Tuesday, July 29 at 12:30 p.m.
b. Ennis, Montana.
c. US Air Force personnel
d. 30 minutes.
e. Two to five flat disk-shaped objects were observed. One disk hovered for 3–4 minutes while the others circled it.

**105.** a. Wednesday, July 30 at 11:02 p.m.
b. Albuquerque, New Mexico.
c. US Air Force First Lieutenant George Funk.

d. 10 minutes.

e. One orange light in the sky remained stationary for the duration of the sighting.

**106.** a. Saturday, August 2 at 3:00 a.m.

b. Lake Charles, Louisiana.

c. US Air Force First Lieutenant W.A. Theil and one enlisted man.

d. 3–4 seconds.

e. One red ball with a blue flame tail flew straight and level for the duration of the sighting.

**107.** a. Monday, August 4 at 2:20 a.m.

b. Phoenix, Arizona.

c. US Air Force Airman Third Class W. F. Vain.

d. 5 minutes.

e. One yellow ball lengthened and narrowed to a plate shape and flew straight and level for the duration of the sighting.

**108.** a. Tuesday, August 5 at 11:30 p.m.

b. Haneda Air Force Base, Japan.

c. US Air Force F-94 jet interceptor pilots First Lieutenant W. R. Holder and First Lieutenant A. M. Jones, and the Haneda Air Force Base control tower operators.

d. 50–60 minutes.

e. Airborne radar tracked an unknown target for 90 seconds. Control tower operators then watched for 50–60 minutes while a dark shape with a visible light flew as fast as 330 knots (380 m.p.h.), hovered, flew curves, and performed a variety of impossible maneuvers.

**109.** a. Saturday, August 9 at 10:50 a.m.

b. Lake Charles, Louisiana.

c. US Air Force Airman Third Class J. P. Raley.
d. 5–6 minutes.
e. One disk-shaped object flew very fast and then hovered for 2 seconds during the sighting.

**110.** a. Wednesday, August 13 at 9:45 p.m.
b. Tokyo, Japan.
c. US Air Force Marine Corps pilot Major D. McGough.
d. 2–3 minutes.
e. One orange light flew a left orbit at 8,000 feet and 230 m.p.h., spiraled down to no more than 1,500 feet, remained stationary for 2–3 minutes and then went out. An attempted interception was unsuccessful.

**111.** a. Tuesday, August 19 at 2:38 p.m.
b. Red Bluff, California.
c. Ground Observer Corpsman Albert Lathrop.
d. 25 seconds.
e. Two objects, shaped like fat bullets, flew straight and level and very fast for the duration of the sighting.

**112.** a. Thursday, August 21 at 11:54 p.m.
b. Dallas, Texas.
c. Ex-artillery observer Jack Rossen.
d. 1½ minutes.
e. Three blue-white lights hovered, then descended; towards the end of the sighting one of them descended more.

**113.** a. Saturday, August 23 at 4:10 a.m.
b. Akron, Ohio.
c. US Air Force Second Lieutenant H. K. Funseth, a ground radar observer, and two US Navy men.

d. 7 minutes.
e. One pulsing amber light was seen to fly straight and level for the duration of the sighting.

114. a. Sunday, August 24 at 10:15 a.m.
b. Hermanas, Mexico.
c. Georgia Air National Guard F-84G jet fighter pilot Colonel G. W. Johnson.
d. 10 minutes.
e. Two 6-foot silver balls flew in abreast formation; one turned grey rapidly, the other changed color slowly. One of the silver balls changed to a long grey shape during a turn.

115. a. Tuesday, August 26 at 12:10 a.m.
b. Lathrop Wells, Nevada.
c. US Air Force Captain D. A. Woods.
d. Brief.
e. One large, round, very bright object with a V-shaped contrail that had a dark cone in the center, flew very fast, hovered, made an instantaneous 90 degree turn, which was followed by a gentle climb and a final sudden acceleration.

116. a. Thursday, August 28 at 9:30 p.m.
b. Chickasaw and Brookley Air Force Base, Alabama.
c. US Air Force control tower operators, an officer from the US Air Force Office of Special Investigations, and several others.
d. 1 hour, 15 minutes.
e. Six objects, varying from fiery red to sparkling diamond appearance, hovered and flew erratically up and down for the duration of the sighting.

**117.** a. Friday, August 29 at 10:50 a.m.
b. West of Thule, Greenland.
c. Two US Navy pilots flying a P4Y-2 patrol plane.
d. 2–3 minutes.
e. Three white disk-shaped or spherical objects hovered, then flew very fast in a triangular formation.

**118.** a. Monday, September 1 at 10:30 p.m.
b. Marietta, Georgia.
c. Ex-artillery officer Mr. Bowman and twenty-four others.
d. 15 minutes.
e. A red, white, and blue-green object spun and shot off sparks for the duration of the sighting.

**119.** a. Monday, September 1 at 10:50 p.m.
b. Marietta, Georgia.
c. An ex-Army Air Force B-25 gunner.
d. Brief.
e. Two large white disk-shaped objects with green vapor trails flew in trail formation, merged, and then flew away very fast.

**120.** a. Monday, September 1 at 4:45 a.m.
b. Yaak, Montana.
c. Visual sighting by two US Air Force enlisted men; radar tracking seen by three men using an AN/FPS-3 radar set.
d. 1 hour.
e. Two small, varicolored lights became black silhouettes at dawn and flew erratically for the duration of the sighting.

**121.** a. Saturday, September 6 at 1:30 a.m.
b. Lake Charles Air Force Base, Louisiana.

c. Technical Sergeant J. E. Wilson and two enlisted men.
d. 2 hours.
e. One bright star-like light moved about the sky for the duration of the sighting.

**122.** a. Saturday, September 6 at 4:55 p.m.
b. Tucson, Arizona.
c. Ex-Congresswoman Mrs. Isabella King and Bill McClain.
d. 1½ minutes.
e. One orange, teardrop-shaped object whirled on its vertical axis, descended very fast, stopped, and then retraced its path upwards while whirling in the opposite direction.

**123.** a. Tuesday, September 9 at 9:00 p.m.
b. Rabat, French Morocco.
c. E. J. Colisimo, a civilian illustrator with US Air Force Intelligence.
d. 5 seconds.
e. One disk with lights along part of its circumference flew twice as fast as a T-33 jet trainer in a slightly curved path for the duration of the sighting.

**124.** a. Sunday, September 14 at 8:40 p.m.
b. Santa Barbara, California.
c. US Air Force C-54 transport pilot Tarbutton.
d. 30 seconds.
e. One blue-white light traveled straight and level, then went up.

**125.** a. Sunday, September 14; time unknown.
b. In the North Atlantic, between Ireland and Iceland.
c. Military personnel from several countries aboard ships in the NATO "Operation Mainbrace" exercise.

d. Brief.
e. One blue-green triangle was observed flying 1,500 m.p.h.; three objects in a triangular formation gave off white light exhaust at 1,500 m.p.h.

**126.** a. Monday, September 16 at 6:22 p.m.
b. Portland, Maine.
c. The crew of a US Navy P2V Neptune patrol plane, visually and by radar.
d. 20 minutes.
e. A group of five lights was seen at the same time a long, thin blip was being tracked on radar.

**127.** a. Monday, September 16 at 7:30 p.m.
b. Warner-Robbins Air Force Base, Georgia.
c. Three US Air Force officers and two civilians.
d. 15 minutes.
e. Two white lights flew abreast at 100 m.p.h. for the duration of the sighting.

**128.** a. Monday, September 23, no time given.
b. Gander Lake, Newfoundland, Canada.
c. A Pepperell Air Force Base operations officer and seven other campers.
d. 10 minutes.
e. One bright white light, which reflected on the lake, flew straight and level at 100 m.p.h. for the duration of the sighting.

**129.** a. Tuesday, September 24 at 3:30 p.m.
b. Charleston, West Virginia.
c. The crew of a US Air Force B-29 bomber.
d. 15 minutes.

e. Many bright, metallic particles or flashes, up to 3 feet in length, streamed past the B-29 for the duration of the sighting.

**130.** a. Thursday, September 26 at 11:16 p.m.
b. 400 miles north-northwest of the Azores Islands.
c. The pilot, copilot, engineer and aircraft commander of a US Air Force C-124 transport plane.
d. Over 1 hour.
e. Two distinct green lights were seen to the right and slightly above the C-124 and at one time seemed to turn toward it. The lights alternated leading each other during the duration of the sighting.

**131.** a. Sunday, September 29 at 8:15 p.m.
b. Southern Pines, North Carolina.
c. US Army Reserve First Lieutenant C. H. Stevens and two others.
d. 15 minutes.
e. One green ellipse with a long tail orbited for the duration of the sighting.

**132.** a. Sunday, September 29 at 3:15 p.m.
b. Aurora, Colorado.
c. US Air Force Technical Sergeant B. R. Hughes.
d. 5–6 minutes.
e. Five or six circular objects, bright white but not shiny, circled in trail formation for the duration of the sighting.

**133.** a. Wednesday, October 1 at 6:57 p.m.
b. Shaw Air Force Base, South Carolina.
c. RF-8O reconnaissance jet pilot US Air Force First Lieutenant T. J. Pointek.
d. 23 minutes.

e. One bright white light flew straight, then vertical, then hovered, and then made an abrupt turn during an attempted intercept.

**134.** a. Tuesday, October 7 at 8:30 p.m.
b. Alamogordo, New Mexico.
c. US Air Force Lieutenant Bagnell.
d. 4–5 seconds.
e. One pale blue oval, with its long axis vertical, flew straight and level for the duration of the observation, covering 30 degrees during the sighting.

**135.** a. Friday, October 10 at 6:30 p.m.
b. Otis Air Force Base, Massachusetts.
c. A US Air Force Staff Sergeant and two other enlisted men.
d. 20 minutes.
e. One blinking white light moved like a pendulum for the duration of the sighting and then shot straight up.

**136.** a. Friday, October 17 at 9:15 p.m.
b. Taos, New Mexico.
c. Four US Air Force officers.
d. 2–3 seconds.
e. One round, bright blue light moved from north to northeast at an elevation of 45 degrees for the duration of the sighting and then burned out.

**137.** a. Friday, October 17 at 11:00 p.m.
b. Tierra Amarilla, New Mexico.
c. One unidentified military person.
d. 20 seconds.
e. One white streamer moved in an arc at an estimated 3,000 m.p.h. for the duration of the sighting.

**138.** a. Sunday, October 19 at 1:30 p.m.
b. San Antonio, Texas.
c. Ex-US Air Force air crewman Woolsey.
d. 3–4 minutes.
e. Three circular aluminum objects, one of which was olive-drab colored on the side, flew in a rough V-formation. One object flipped slowly and another object stopped during the sighting.

**139.** a. Sunday, October 19 at 6:58 p.m.
b. 500 miles south of Hawaii.
c. The crew of a US Air Force C-50 transport plane.
d. 20 seconds.
e. One round yellow light with a red glowing edge, estimated at 100 feet in diameter, flew at 300–400 knots (350–450 m.p.h.) for the duration of the sighting.

**140.** a. Friday, October 24 at 8:26 p.m.
b. Elberton, Alabama.
c. US Air Force Lieutenant Rau and Captain Marcinko flying a Beech T-ll trainer.
d. 5 seconds.
e. One object, shaped like a plate with a brilliant front and a vague trail, flew with its concave surface forward for the duration of the sighting.

**141.** a. Wednesday, October 29 at 7:50 a.m.
b. Erding Air Depot, West Germany.
c. US Air Force Staff Sergeant Anderson, Airman Second Class Max Handy.
d. 20 seconds.
e. One round object silhouetted against a cloud flew straight and level and smooth at 400 m.p.h. for the duration of the sighting.

**142.** a. Friday, October 31 at 7:40 p.m.
b. Fayetteville, Georgia.
c. US Air Force Lieutenant James Allen.
d. 1 minute.
e. An orange, blimp-shaped object, 80 feet long and 20 feet high, flew at treetop level, crossed over Lieutenant Allen's car, and then climbed out at 45 feet altitude at tremendous speed at the end of the sighting. When the object passed over Allen's car, his car radio stopped playing.

**143.** a. Monday, November 3 at 6:29 p.m.
b. Laredo Air Force Base, Texas.
c. Two control tower operators, including one named Lemaster.
d. 3–4 seconds.
e. One long, elliptical, white-grey light flew very fast, paused, and then increased speed during the observation.

**144.** a. Thursday, November 13 at 2:20 a.m.
b. Opheim, Montana.
c. Radar tracking personnel at the US Air Force 779th AC&W station.
d. 1 hour, 28 minutes.
e. An unexplained radar track was followed for the duration of the sighting at 158,000 feet altitude (30 miles) and at a speed of 240 m.p.h. Radar used was FPS/3 (PPI).

**145.** a. Thursday, November 13 at 2:43 a.m.
b. Glasgow, Montana.
c. US Weather Bureau observer Earl Oksendahl.
d. 20 seconds.
e. Five oval-shaped objects with lights all around them flew in a V-formation for the duration of the sighting. Each object

seemed to change position vertically by climbing or diving as if to hold formation. The formation came from the northwest, made a 90 degree turn overhead, and then flew away to the southwest.

**146.** a. Saturday, November 15 at 7:02 a.m.
b. Wichita, Kansas.
c. US Air Force Major R. L. Wallander, Captain Belleman, Airman Third Class Phipps.
d. 3–5 minutes.
e. One orange object varied in shape as it made jerky upward sweeps with 10–15 second pauses during the duration of the sighting.

**147.** a. Thursday, November 27 at 12:10 p.m.
b. Albuquerque, New Mexico.
c. The pilot and crew chief of a US Army Air Force B-26 bomber.
d. 20 minutes.
e. A series of black smoke bursts similar to antiaircraft fire was seen visually for the duration of the sighting, and then tracked on radar until approximately 12:30 a.m.

**148.** a. Monday, December 8 at 8:16 p.m.
b. Ladd Air Force Base, Alaska.
c. Pilot First Lieutenant D. Dickman and radar operator First Lieutenant T. Davies in a US Air Force F-94 jet interceptor (serial number 49-2522).
d. 10 minutes.
e. One white, oval light (which changed to red at higher altitude), flew straight and level for 2 minutes, then climbed at phenomenal speed on an erratic flight path for the remainder of the sighting.

**149.** a. Tuesday, December 9 at 5:45 p.m.
b. Madison, Wisconsin.
c. Captain Bridges and First Lieutenant Johnson in a US Air Force T-33 jet trainer.
d. 10 minutes.
e. Four bright lights, in diamond formation, flew at 400 m.p.h. and were passed by the T-33 at 450 m.p.h. during the sighting.

## 1953

**150.** a. Thursday, January 8 at 7:15 a.m.
b. Larson Air Force Base, Washington.
c. Men from the 82nd Fighter-Interceptor Squadron, including the squadron commander. All were on the ground.
d. 15 minutes.
e. One green, disk-shaped or round object flew southwest below the clouds with a vertically bobbing motion and sideways movements for the duration of the sighting.

**151.** a. Saturday, January 10 at 3:45 p.m. or 4:00 p.m.
b. Sonoma, California.
c. Retired Colonel Robert McNab and Mr. Hunter of the Federal Security Agency.
d. 60–75 seconds.
e. One flat object shaped like a pinhead made three 360 degree right turns in 9 seconds, made abrupt 90 degree turns to the right and left, stopped, accelerated to its original speed and finally flew out of sight vertically at the conclusion of the sighting.

**152.** a. Wednesday, January 28 at 6:05 p.m.
b. Corona, California.

c. US Air Force Technical Sergeant George Beyer.
d. 12 minutes.
e. Five 25 foot green spheres flew in a V-formation, then changed to trail formation at which time the end objects turned red.

**153.** a. Tuesday, February 17 at 10:04 p.m.
b. Port Austin, Michigan.
c. Two officers and three airmen of the US Air Force AC&W squadron, visually and by radar.
d. 22 minutes.
e. The visual object appeared to be larger and brighter than a star and changed color. It was seen to move slowly for 5 minutes until 10:09 p.m. Radar picked up a target at 10:08 p.m. moving in a similar direction for 17 minutes, at similar speed.

**154.** a. Friday, February 20, from early evening until 10:30 p.m.
b. Pittsburg-Stockton, California.
c. US Air Force B-25 bomber pilots.
d. 8 minutes plus.
e. Sighting number one was a bright yellow light seen for 8 minutes. Sighting number two was a bright light which flew on a collision course, dimmed, and then climbed away fast.

**155.** a. Friday, February 27 at 11:58 a.m.
b. Shreveport, Louisiana.
c. A US Air Force airman who was also a private pilot.
d. 4 minutes.
e. Five yellow disks made circular turns, fluttered, and then three of them vanished. The other two then flew erratic square turns for the remainder of the sighting.

**156.** a. Saturday, March 14 at 11:45 p.m.
b. North of Hiroshima, Japan.
c. Radar and visual observation made by ten crew members of a US Navy P2V-5 patrol plane.
d. 5 minutes.
e. Several groups of five to ten colored lights, totaling ninety to one hundred lights, slowly moved aft off the left side of the airplane. The objects were detected visually and by airborne radar for the duration of the sighting.

**157.** a. Wednesday, March 25 at 3:05 p.m.
b. San Antonio, Texas.
c. US Air Force Captain D. E. Cox and Mrs. Cox.
d. 1½ hours.
e. Several lights were visually observed for the period of the sighting. Some of the lights moved straight, others made 360 degree turns while being observed.

**158.** a. Friday, March 27 at 7:25 p.m.
b. Mount Taylor, New Mexico.
c. 4 minutes.
d. The pilot of a US Air Force F-86 jet fighter flying at 600 knots (700 m.p.h.).
e. One bright orange circle flew at 800 knots (900 m.p.h.) and executed three fast rolls. The pilot chased the object for the duration of the sighting.

**159.** a. Wednesday, April 8 at 7:55 p.m.
b. Fukuoka, Japan.
c. First Lieutenant D. J. Pichon, the pilot of a US Air Force F-94B jet interceptor.
d. 45 seconds.

e. One bright blue light descended, accelerated, flew parallel to the F-94, increased its speed and then blinked out at the conclusion of the sighting.

**160.** a. Wednesday, April 15 at 5:45 p.m.
b. Tucson, Arizona.
c. Staff Sergeant V. A. Locey.
d. Approximately 3 minutes, 40 seconds.
e. Three orange lights were seen three times: once for 3 minutes, once for 30 seconds, and once for a few seconds.

**161.** a. Friday, May 1 at 11:35 p.m.
b. Goose Air Force Base, Labrador, Canada.
c. The pilot and radar operator of a US Air Force F-94 jet interceptor, and a control tower operator.
d. 30 minutes.
e. One white light evaded an interception attempt by the F-94 for the entire duration of the sighting.

**162.** a. Monday, June 22 at 2:10 a.m.
b. Goose Air Force Base, Labrador, Canada.
c. The pilot and radar operator of a US Air Force F-94 jet interceptor.
d. 5 minutes.
e. One red light, flying at an estimated 1,000 knots (1,100 m.p.h.), eluded the chasing F-94 for the duration of the sighting.

**163.** a. Wednesday, June 24 at 11:30 p.m.
b. Iwo Jima, Bonin Islands.
c. The crew of a US Air Force KB-29 aerial tanker plane.
d. 2 minutes.

e. Radar tracked an unidentified target which twice approached to within 6 miles of the airplane and once to within 1/2 mile of the plane.

**164.** a. Thursday, August 20 at 9:05 p.m.
b. Near Castle Air Force Base, California.
c. The crew of a TB-29 bomber/trainer plane.
d. Brief.
e. One greyish oval object made four passes at the airplane (three times at 10–20 miles distance), then dived vertically near the plane, appearing visually as if had become two objects.

**165.** a. Thursday, August 27 at 9:45 p.m.
b. Greenville, Mississippi.
c. A US Air Force pilot, a Master Sergeant, and other military personnel, all on the ground.
d. 50 minutes.
e. One airborne meandering light was observed for the duration of the sighting.

**166.** a. Thursday, December 24 at 8:04 a.m.
b. El Cajon, California.
c. US Navy Lieutenants J. B. Howard and L. D. Linhard, flying F9F-2 jet fighters.
d. 5 minutes.
e. Ten silver, oval objects flew at more than 400 knots (450 m.p.h.), straight and level, for the duration of the sighting.

## 1954

**167.** a. Friday, March 5 at 8:00 p.m.
b. Nouasseur, French Morocco.
c. The crews of more than one US Air Force KC-97 aerial tanker planes.
d. Unknown, but probably brief.
e. One unidentified object or light made intermittent passes at the KC-97s; another flew straight and level nearby.

**168.** a. Friday, March 12 at 9:35 a.m.
b. Nouasseur, French Morocco.
c. US Air Force First Lieutenant Robert Johnson flying an F-86 jet fighter.
d. 30 seconds.
e. Johnson chased an object at more than 530 m.p.h. for the duration of the sighting, but was unable to catch it. The object appeared to be the size of a fighter plane but had neither tanks nor trails.

**169.** a. Tuesday, May 11 at 10:45 p.m.
b. Washington, D.C.
c. Three US Air Force air police officers at Washington National Airport.
d. 2 minutes, 15 seconds.
e. Two bright lights were seen on three occasions to fly straight and level, make 90 degree turns and then fade away. Each sighting lasted about 45 seconds.

**170.** a. Tuesday, June 1 at 9:00 p.m.
b. From 400 miles south of Minneapolis, Minnesota to the environs of that city.

c. The crew of a US Air Force B-47 jet bomber flying at an altitude of 34,000 feet.
d. 1 hour.
e. One object with running lights flew at 24–44,000 feet altitude for the duration of the sighting.

**171.** a. Thursday, June 10 at 9:09 p.m.
b. Estacado, Texas.
c. US Air Force pilot Captain Bill McDonald, in flight.
d. 30 seconds.
e. One white light descended at 45 degrees from a great altitude, passed under McDonald's aircraft, made two 360 degree turns, and then went out at the conclusion of the sighting.

**172.** a. Tuesday, June 22 at 9:00 p.m.
b. Miami Beach, Florida.
c. US Marine Corps Majors E. Buchser and J. V. Wilkins.
d. 7 minutes.
e. One meteor-like object descended, stopped, and then became extremely bright.

**173.** a. Monday, August 2 at 5:17 p.m.
b. Westlake, Ohio.
c. Ex-Army Air Force B-17 gunner (19 missions) N. E. Schroeder.
d. 20 seconds.
e. One thin, bright ellipse that looked like polished metal hovered for 5–8 seconds, dropped down 3,000 feet in 3 seconds, hovered again, and faded out after a total of 20 seconds in view.

**174.** a. Sunday, August 15 at 10:20 p.m.
b. San Marcos, Texas.

c. US Air Force Major W. J. Davis and Captain R. D. Sauers flying a C-47 transport plane.
d. 5 minutes.
e. One dark blue oblong object paced the C-47, veered away, then crossed in front of it.

175. a. Tuesday, September 21 at 1:00 a.m.
b. Barstow, California.
c. Two local police officers, four US Marine Corps police officers, and one highway patrolman.
d. 20 minutes.
e. A red-orange ball giving off sparks and a smaller light made a zigzag descent and then hovered.

176. a. Thursday, October 28 at 5:32 p.m.
b. Miho Air Base, Japan.
c. US Air Force pilots Lieutenant Colonel O. C. Cook and Lieutenant J. W. Brown on the ground using 7x50 binoculars.
d. 45 seconds
e. One brilliant white, round-oval object climbed in front of clouds, brightened, and then turned 90 degrees to the north.

## 1955

177. a. Saturday, January 1 at 6:44 a.m.
b. Cochise, New Mexico.
c. An instructor and student pilot in a US Air Force B-25 bomber/trainer.
d. 5–7 minutes.

e. A metallic disk, 120–130 feet in diameter and shaped like two pie pans face-to-face, paced the B-25, showing both its edge and its face, for the duration of the sighting.

**178.** a. Tuesday, February 1 at 7:55 p.m.
b. 20 miles east of Cochise, New Mexico.
c. Instructor Captain D. F. Ritzdorf and aviation cadet F. W. Miller in a TB-25 bomber/trainer.
d. 8 minutes.
e. One red and white ball hovered off the left wing of the TB-25 for 5 minutes and then made a very fast climb, remaining in view for an additional 3 minutes.

**179.** a. Wednesday, February 2 at 11:50 a.m.
b. Miramar Naval Air Station, California.
c. US Navy Commander J. L. Ingersoll.
d. Brief.
e. One highly polished sphere with reddish-brown coloring fell, then instantly accelerated to 1,000–1,500 m.p.h.

**180.** a. Saturday, April 30 at 7:30 a.m.
b. Travis County, Texas.
c. US Air Force Wing Intelligence Officer Major L. J. Pagozalski.
d. 2–3 seconds.
e. Four black objects in a cluster made a whooshing sound like a zephyr.

**181.** a. Wednesday, May 4 at 12:38 p.m.
b. Keflavik, Iceland.
c. Lieutenant Colonel E. J. Stealy and First Lieutenant J. W. Burt.
d. 5–8 seconds.

e. About ten round, white objects, one of which left a brief smoke trail, flew in an irregular formation, some of them making erratic movements during the sighting.

**182.** a. Monday, May 23 at midnight.
b. Cheyenne, Wyoming.
c. US Air Force Airmen I. J. Shapiro and E. C. Ingber.
d. 5 minutes.
e. Two slender, vertical rectangles were seen low on the horizon, and two ovals with tops (the ovals appeared dark with dark blue illumination) flew higher.

**183.** a. Thursday, August 11 at 11:45 a.m.
b. Iceland.
c. Second Lieutenant E.J. Marlow.
d. 3–4 minutes.
e. Twelve grey objects, from cigar to egg-shaped, varied their formation from elliptical to wavy line to scattered to straight line to trail formation. Their speed varied from hover to 1,000 m.p.h.

**184.** a. Sunday, November 20 at 5:20 p.m.
b. Lake City, Tennessee.
c. Operations Officer Captain B. G. Denkler and five men of the US Air Force 663rd AC&W Squadron, as well as many others in the vicinity during the sighting.
d. 4–15 minutes.
e. Two oblong, bright orange, semi-transparent objects flew erratically at terrific speed toward and away from each other.

**185.** a. Friday, November 25 at 10:30 a.m.
b. La Veta, Colorado.

c. State Senator S. T. Taylor.
d. 5 seconds.
e. One luminous, green-blue, jelly-like, dirigible-shaped object with a fat front that tapered toward the tail appeared overhead diving at a 45 degree angle. The object then reduced its angle to 30 degrees.

## 1956

**186.** a. Sunday, February 12 at 11:25 p.m.
b. Goose Bay, Labrador, Canada.
c. F-89 pilot Bowen and radar observer Crawford.
d. 1 minute.
e. One green and red object rapidly circled the aircraft while being tracked on radar during the sighting.

**187.** a. Wednesday, April 4 at 3:15 p.m.
b. McKinney, Texas.
c. Captain Roy Hall, US Army retired; Charles Anderson; and others.
d. 6 hours.
e. One fat, oblong object with two lines around its middle remained stationary for the duration of the sighting. Some observed the object through a 6-inch telescope, others used a 55–200x telescope.

**188.** a. Tuesday, September 4 at 9:00 p.m.
b. Dallas, Texas.
c. US Marine Corps Technical Sergeant R. D. Rogers and his family.
d. 23 minutes.
e. One large star changed to a red color while remaining stationary for 20 minutes. The object then flew west at 200 knots (230 m.p.h.) during the remaining 3 minutes of the sighting.

**189.** a. Thursday, November 1 at 5:30 p.m.
b. 60 miles east of St. Louis, Missouri, in Illinois.
c. Intelligence Division Chief (Aerial Weather Reconnaissance Officer) US Air Force Captain W. M. Lyons flying a T-33 jet trainer.
d. 2 minutes.
e. One orange light with a blue tinge flew across the sky for the duration of the sighting.

**190.** a. Friday, November 30 at 12:48 p.m.
b. Charleston Air Force Base, South Carolina.
c. US Air Force aerial navigator Major D. D. Grimes.
d. 10 minutes.
e. One unspecified and unidentified object flew at an estimated 100-foot altitude over water for the duration of the sighting.

**191.** a. Monday, December 31 at 2:10 a.m.
b. Guam.
c. US Air Force First Lieutenant Ted Brunson flying an F-86D jet interceptor.
d. Brief.
e. One round, white object flew under the F-86D. The jet was unable to turn as sharply as the object did.

## 1957

**192.** a. Friday, September 20 at 8:00 p.m.
b. Kadena Air Force Base, Okinawa.
c. Staff Sergeant H. T. O'Connor and Staff Sergeant H. D. Bridgeman.
d. 15 minutes.

e. One translucent and fluorescent object shaped like a Coke bottle without the neck made four 5–10 second passes from north to south, with 4–5 minutes between passes.

**193.** a. Tuesday, October 8 at 9:17 a.m.
b. Seattle, Washington.
c. Two US Army sergeants.
d. 25–30 seconds.
e. Two flat, round, white objects flew in trail formation along an irregular path, frequently banking during the duration of the sighting.

**194.** a. Saturday, November 30 at 2:11 p.m.
b. New Orleans, Louisiana.
c. Three US Coast Guardsmen.
d. 20 minutes.
e. One round object turned white, then gold, then separated into three parts and turned red.

## 1958

**195.** a. Monday, April 14 at 1:00 p.m.
b. Lynchburg, Virginia.
c. US Air Force Major D. G. Tilley flying a C-47 transport.
d. 4 seconds.
e. One grey-black rectangular object rotated very slowly on its horizontal axis for the duration of the sighting.

**196.** a. Friday, June 20 at 11:05 p.m.
b. Ft. Bragg, North Carolina.
c. Battalion Communication Chief Seaman First Class A. Parsley.

d. 10 minutes.

e. One silver, circular object hovered, then oscillated slightly, then moved away at great speed. The lower portion of the object was seen through a green haze.

**197.** a. Monday, November 3 at 2:01 p.m.

b. Minot, North Dakota.

c. Master Sergeant and medic William R. Butler.

d. 1 minute.

e. One bright green object shaped like a dime exploded. A second object, which was smaller, round, and silver, moved toward the location of the first object at high speed after the first object exploded.

## 1959

**198.** a. Tuesday, June 30 at 8:23 p.m.

b. Patuxent River Naval Air Station, Maryland.

c. US Navy Commander D. Connolly.

d. 20–30 seconds.

e. One gold, oblate-shaped, metallic object with sharp edges, nine times as wide as it was thick, flew straight and level for the duration of the sighting.

**199.** a. Sunday, October 4 at 9:25 p.m.

b. Quezon, Philippine Islands.

c. US Navy Lieutenant C. H. Pogson, Chief Petty Officer K. J. Moore.

d. 15 minutes.

e. One large round or oval object changed color from red to red-orange and flew straight and level for the duration of the sighting.

**200.** a. Tuesday, October 6 at 8:15 p.m.

b. Lincoln, Nebraska.

c. Lieutenant Colonel L. Liggett of the Selective Service and Mrs. Liggett.

d. 2 minutes.

e. One round, white-yellow light made several abrupt turns and flew very fast for the duration of the sighting.

**201.** a. Monday, October 19 at 9:25 p.m.

b. Plainville, Kansas.

c. US Air Force Academy engineering instructor Captain F. A. Henney, flying a T-33 jet trainer.

d. 30 seconds.

e. One bright yellowish light came head-on at the T-33. The pilot avoided it and the light subsequently dimmed.

## 1960

**202.** a. Sunday, April 17 at 8:29 p.m.

b. Richards-Gebauer Air Force Base, Missouri.

c. US Air Force Major J.G. Ford and Link representative A. Chapdelaine, using a 48x telescope.

d. 2½ minutes.

e. One reddish glow made an odd orbit for the duration of the sighting.

**203.** a. Tuesday, November 29 at 6:38 p.m.

b. South of Kyushu, Japan.

c. US Air Force Lieutenant Colonel R. L. Blewlin and Major F. B. Brown, flying a T-33 jet trainer.

d. 10 minutes.

e. One white light slowed and paralleled the course of the T-33 for the duration of the sighting.

## 1961

**204.** a. Monday, April 24 at 3:34 a.m.
b. 200 miles southwest of San Francisco, California.
c. Aircraft commander Captain H. J. Savoy and navigator First Lieutenant M. W. Rand, on a US Air Force RC-121D patrol plane.
d. 8 minutes.
e. One reddish-white, round object or light was observed for the duration of the sighting.

**205.** a. Friday, June 2 at 10:17 p.m.
b. Miyako Jima, Japan.
c. First Lieutenant R. N. Monahan and Hazeltine Electric Co. technical representative D. W. Mattison.
d. 5 minutes.
e. One blue-white light flew an erratic course at varying speed in an arc-like path for the duration of the sighting.

**206.** a. Wednesday, December 13 at 5:05 p.m.
b. Washington, D.C.
c. Ex-US Navy pilot W. J. Myers and others.
d. 1–3 minutes.
e. One dark diamond-shaped object with a bright tip flew straight and level for the duration of the sighting.

## 1962

**207.** a. Monday, March 26 at 1:35 p.m.
b. Ramstein Air Base, West Germany.
c. US Air Force Captain J. M. Lowery, from an unspecified aircraft.
d. 5–8 seconds.
e. One thin, cylindrical object that was proportioned approximately one-third snout and two-thirds tail fins flew at an estimated Mach 2.7 (2,000 m.p.h.) for the duration of the sighting.

**208.** a. Thursday, June 21 at 4:00 a.m.
b. Indianapolis, Indiana.
c. Lieutenant Colonel H. King and tail gunner Master Sergeant Roberts, aboard a B-52 heavy jet bomber.
d. 3 minutes.
e. Three bright, star-like lights were seen. Ten seconds later, two more were seen.

## 1963

**209.** a. Saturday, September 14 at 3:15 p.m.
b. Susanville, California.
c. Retired US Forest Service lookout instructor E. A. Grant.
d. 10 minutes.
e. One round object intercepted a long object and either attached itself to the latter or disappeared.

## 1964

**210.** a. Saturday, May 9 at 10:20 p.m.
b. Chicago, Illinois.
c. US District Court reporter J. R. Betz.
d. 3 seconds.
e. Three light green crescent-shaped objects, about half the apparent size of the Moon, flew very fast in tight formation from east to west, oscillating in size and color for the duration of the sighting.

**211.** a. Tuesday, May 26 at 7:43 p.m.
b. Cambridge, Massachusetts.
c. Royal Air Force pilot and ex-Smithsonian satellite tracker P. Wankowicz.
d. 3–4 seconds.
e. One thin, white ellipsoid, 3½ times as long as it was wide, flew straight and level for the duration of the sighting.

**212.** a. Monday, August 10 at 5:16 a.m.
b. Wake Island.
c. Aircraft commander Captain B. C. Jones and navigator First Lieutenant H. J. Cavender in a parked US Air Force C-124 transport plane.
d. 2 minutes.
e. One reddish, blinking light approached the runway, stopped and made several reverses during the sighting.

**213.** a. Tuesday, August 18 at 12:35 a.m.
b. Over the Atlantic Ocean, 200 miles east of Dover, Delaware.
c. Major D. W. Thompson and First Pilot First Lieutenant J. F. Jonke on a US Air Force C-124 transport plane.

d. 2 minutes.

e. One round, blurred, reddish-white object was on a collision course with the C-124 from ahead and below. The airplane evaded the object.

## 1965

**214.** a. Sunday, April 4 at 4:05 a.m.

b. Keesler Air Force Base, Mississippi.

c. US Air Force weather observer Airman Second Class Corum.

d. 15 seconds.

e. One 40-foot black, oval object with four lights along the bottom, flew in and out of the clouds for the duration of the sighting.

## 1966

**215.** a. Sunday, March 20 at 12:15 a.m.

b. Miami, Florida.

c. US Air Force Reserve Major K. C. Smith and an employee of NASA at Cape Kennedy.

d. 5 minutes.

e. One pulsating light which varied from white to intense blue made a jerky ascent and then rapidly accelerated away to the north.

**216.** a. Monday, June 27 at 4:00 a.m.

b. 400 miles east of Wake Island.

c. Radio Officer Steffen Soresen of the ship *Mt. Vernon Victory*.

d. Several minutes.

e. One "cloud" expanded with a light inside, and then accelerated away.

**217.** a. Friday, August 19 at 4:50 p.m.
b. Donnybrook, North Dakota.
c. US Border Patrolman Don Flickenger.
d. 5 minutes.
e. A round disk with a domed top, 30 feet in diameter and 15 feet high, colored white, silvery or aluminum, moved across a valley from the southeast, hovered over a reservoir, appeared to land in a small field, then rose up into clouds very rapidly.

## 1967

**218.** a. Monday, February 20 at 3:10 a.m.
b. Oxford, Wisconsin.
c. US Air Force veteran/truck driver Stanton Summer.
d. 2 minutes.
e. One orange-red object flew parallel to Summer's truck for the duration of the sighting.

# TRUTH

## Montana's Malmstrom Air Force Base

At Montana's Malmstrom Air Force Base in 1967, Captain Robert Salas, a missile launch control officer, reported a complete shutdown of the missile launch system in multiple silos after a bright light was observed by Air Force personnel hovering over the base. The missile controls were hardwired in the silos, thus making any interruption of the signal without a physical contact very mysterious. The Air Force investigators reported that the missile launch controls had suffered an unknown malfunction, but did not mention the UFO sighting.

### The Last Word?

*The key was to find a place that was accessible yet not too inviting.*

—Major David J. Shea talking in 1969 about where to store the voluminous Project Blue Book files after the project was shut down.

## Part II

# IN PLAIN SIGHT: UFOS IN THE SKY FROM ANCIENT TO MODERN TIMES

# Astonishing UFO Sightings from 1860–1998

*The single most important event in the history of the human race would be contact with extraterrestrials. Think about that. While the discovery of fire, the invention of language, numbers, and the wheel are undeniably significant in the millennia-long rise of civilization, those events only bring us nearer to the central question of life—they don't answer it. Among everything that occurs on earth every day, the event that would most profoundly change our view of reality and answer our overwhelming search for meaning would be contact with ETs. That would let us know exactly what and who we are in the cosmos.*

*That contact has already occurred, although many people don't believe this to be the case. There are aliens on our planet; they have arrived in UFOs. Millions of people all over the world have seen spaceships.*

—Ellen Crystall, Ph.D., *Silent Invasion: The Shocking Discoveries of a UFO Researcher*

These randomly selected UFO sightings are by no means unique in terms of the plethora of equally impressive UFO sightings compiled over the past century. We think you will find these accounts, though, especially interesting and intriguing. Prepare thyself: Weirdness abounds here!

# TRUTH

### UFO Sighting by European Colonists

In 1639, Royal Governor of the Massachusetts Bay colonies John Winthrop wrote about "a sober discreet man," James Everett, and two others who had been rowing a boat in the Muddy River, which flowed through swampland and emptied into a tidal basin in the Charles River, when they saw a great light in the nighttime sky. "When it stood still, it flamed up, and was about three yards square," the governor reported, "when it ran, it was contracted into the figure of a swine." Over the course of two to three hours, the boatmen said that the mysterious light "ran as swift as an arrow" darting back and forth between them and the village of Charlestown, a distance of approximately two miles. "Diverse other credible persons saw the same light, after, about the same place," Winthrop added. Winthrop wrote about the sighting in his *History of the New England Colonies.* He added that "when the strange apparition finally faded away, the three Puritans in the boat were stunned to find themselves one mile upstream—as if the light had transported them there. The men had no memory of their rowing against the tide, although it's possible they could have been carried by the wind or a reverse tidal flow.

## Friday, July 13, 1860

On this Friday the 13th, almost the entire city of Wilmington, Delaware was lit by a pale blue light and people in the city looked up into the skies and reportedly saw a 200-foot-long object flying at approximately 100 feet altitude. In the July 20, 1860 edition of the *Wilmington Tribune*, it was reported that the craft "moved in a straight line without any inclination downwards." The press report also noted that a pitch black cloud of some sort flew in front of the object, and that behind it, spaced 100 feet apart, flew three "very red and glowing balls." A fourth ball then appeared as the 200-foot-long object turned towards the southeast over the Delaware River. The newspaper also reported that the main craft "gave off sparkles in the manner of a rocket." This astonishing one-minute sighting came to a conclusion when the mysterious craft turned toward the east and flew out of view. (*Wilmington, Delaware*)

## Tuesday, April 20, 1897

Alexander Hamilton, a member of the Kansas House of Representatives, filed an affidavit in which he described the following close encounter: Hamilton, his son, and his farmhands all saw a dark red, cigar-shaped 300-foot long craft land in a field near his home. Beneath the object was what appeared to be a glass cabin in which there were six strange beings. The craft then rose off the ground and Hamilton and the others saw that a cow was hanging from a rope that was dangling from the bottom of the craft. The object rose into the darkness and the following day, the calf's legs, head, and skin were found on the grounds of a nearby farm. Hamilton's credibility was impeccable and later he said, "Every time I drop to sleep I see the cursed thing, all its bright lights and hideous people . . . I don't want any more to do with them." Future assessments of the case cast suspicion on Hamilton and there was speculation that he lied as part of a planned hoax. He was discovered to be a member of a group called the Liars' Club. (*LeRoy, Kansas*)

## Wednesday, December 22, 1909

While putting up Christmas decorations, Mr. and Mrs. William Forsythe of 85 Evergreen Street in Providence, Rhode Island both saw a large UFO in the sky above Providence. According to the December 22, 1909 edition of the *Providence Journal*, "Mrs. Forsythe looked out of the windows [and] her attention was attracted by two red lights in the sky which were different from anything she had ever seen before. She called her husband to the window and both watched the strange spectacle. The lights appeared to be covering a course that was varied, now rather close to the earth and then soaring upward, but always making toward the south. They were able to make out an object which appeared to be in front of the lights. . . . It was moving at such high speeds that they could get little more than a superficial view of the object, although what they saw was enough to satisfy them as to the identity of the contrivance. The Forsythes watched until the lights faded out in the haze on the southern horizon. When they arose in the morning, they promptly told all their friends of what they had seen." (*Providence, Rhode Island*)

## Monday, March 20, 1950

While flying near Little Rock, Arkansas, Captain Jack Adams and his First Officer G. W. Anderson both observed a disk-shaped craft fly above their airliner in an arc. Both pilots reported seeing portholes in the sides of the craft. (*Little Rock, Arkansas*)

## Saturday, May 20, 1950

Seymour L. Hess, a scientist specializing in meteorology and astronomy, saw a craft shaped like a disk or a sphere moving across the sky in what he described as "powered" flight. Hess reported the sighting and in his account stated, "I saw the object between 12:15 and 12:20 P.M . . . from the grounds of the Lowell Observatory. It was moving from the southeast to the northwest. It was extremely prominent and showed some size to the naked eye . . . it was not merely a pinpoint. . . . I could see it well enough to be sure it was not an airplane . . . nor a bird. I saw no evidence of exhaust gases nor any markings on the object." (*Flagstaff, Arizona*)

## Saturday, January 20, 1951

As a Mid-Continent Airlines jet approached Iowa's Sioux City Airport, Pilot Captain Lawrence Vinther, First Officer James F. Bachmeier, airport control tower operators, and the passengers on the plane all reported seeing a cigar-shaped object with bright body lights flying in the plane's flight path. The object suddenly circled back and approached the airplane head-on. All the witnesses then agree that the object suddenly reversed direction, paced the plane for a few seconds, and then flew straight up in the air at a high rate of speed before disappearing. (*Sioux City, Iowa*)

## Saturday, July 2, 1952

US Navy Warrant Officer Delbert C. Newhouse and his wife visually observed a group of ten or twelve UFOs at approximately 11:10 a.m. over Tremonton, Utah. After unpacking his camera, Newhouse shot 30 feet (approximately 1200 frames) of 16 mm color footage of the objects. Newhouse described the objects as "shaped like two saucers, one inverted on top of the other." Newhouse used a Bell and Howell Automaster camera loaded with Daylight Kodachrome film. In 1963, the Air Force issued a statement that the Newhouse film showed a flock of seagulls, even though two earlier analyses of the film (one by the Wright-Patterson AFB photo lab and one by the Navy photo lab at Anacostia) both concluded that the objects were not aircraft, balloons, or birds. (*Tremonton, Utah*)

## Friday, July 18, 1952

American Airlines Captain Paul Carpenter reported seeing three UFOs at one time in the skies over Denver, Colorado. The UFOs appeared as speeding lights and Captain Carpenter reported that they appeared to reverse direction while in flight. (*Denver, Colorado*)

## Monday, October 11, 1952

A woman reported seeing a flying disk hover in one spot for 20 minutes over Newport News, Virginia. When military jet interceptors approached the area, the UFO tilted up and shot away at a rapid speed as the woman and the pilots continued to observe it until it was no longer visible. (*Newport News, Virginia*)

## Sunday, October 26, 1958

As Alvin Cohen and Phillip Small approached a bridge on the Loch Raven Dam in Baltimore, Maryland, they both saw a "large, flat sort of egg-shaped object hanging between 100 to 150 feet off the top of the superstructure of the bridge over the lake." When they were approximately 75 feet away from the bridge, their car died and the electrical system failed. They both got out of the car and watched the UFO for thirty to forty seconds before the object made a sound like an explosion, flashed a bright white light, and started to rise vertically. Within five to ten seconds, the UFO had disappeared from their view. This sighting was corroborated by several others in the area who also heard the explosion and saw the bright light. (*Baltimore, Maryland*)

## Saturday, August 13, 1960

Near midnight on August 13, 1960, Charles Carson and Stanley Scott, two California State Police Officers, observed a large lighted object falling from the sky in Corning. They at first thought it was a disabled airliner and believed the falling plane would strike their police cruiser. Fearing for their safety, they fled the car. The two officers then watched as the object dropped to approximately 100 feet above the ground, stopped in mid-air, reversed direction, and then rapidly climbed to an altitude of approximately 500 feet.

They later described the object as round or oblong in shape, surrounded by a bright glow. The craft had red lights on each end and blinking white lights between the red ones.

As Carson and Scott watched, the craft then performed aerial maneuvers the men later described as "unbelievable." The craft also flew back towards the police cruiser and shone a red light on the vehicle several times.

When the object began to fly in an easterly direction, the police officers followed and observed a second craft join with the first, and then both disappeared below the eastern horizon. Carson and Scott observed the craft for a total of 2.25 hours. The crafts were also observed by two Tehama County deputy sheriffs and the county's night jailer. Radar operators at the Air Force Base in nearby Red Bluff, California confirmed the presence of the object on their radar screens. No explanation has ever been given for this sighting.

In his official report, Officer Carson stated, "I served four years with the Air Force, I believe I am familiar with the northern lights, also weather balloons. Officer Scott served as a paratrooper during the Korean conflict. Both of us are aware of the tricks light can play on the eyes during darkness. We were aware of this at the time. Our observations and estimations of speed, size, etc. came from aligning the object with fixed objects on the horizon. I agree we find it difficult to believe what we were watching, but no one will ever convince us that we were witnessing a refraction of light." (*Corning, California*)

## Friday, June 21, 1963

A student witnessed a gray, spherical UFO with a central row of yellow lights in the skies above Chicago, Illinois. The object moved from west to east, then turned sharply, before disappearing into the northern sky. The student also reported hearing what he described as a "sizzling sound" as the craft flew above him. The witness reported the sighting to authorities and also filed a NICAP report form. (*Chicago, Illinois*)

## Tuesday, August 13, 1963

Dr. Richard Turse, a chemist from Princeton, New Jersey vacationing in Honolulu, Hawaii, observed a round, reddish UFO fly overhead from southwest to northeast. The time was approximately 11:30 p.m. local time and Turse reported the sighting to NICAP, stating that the UFO "traveled across the sky at great speed, making two sharp turns at the same speed." (*Honolulu, Hawaii*)

## Monday, March 15, 1965

While on a hunting expedition in the Florida Everglades with his dogs, James Flynn saw a UFO shaped like a broad, upside-down cone hovering about 200 feet above a grove of cypress trees. The object was about 25 feet high and 50 feet wide and had square windows that emitted a yellowish glow. Flynn also saw a reddish glow emanating from the bottom of the craft. When Flynn approached the craft, a beam of light shot out and hit him in the forehead. Flynn blacked out and when he awakened, he was temporarily blind. After his sight partially returned, Flynn saw a burned circle on the ground beneath where the UFO had been hovering. Flynn was hospitalized and sustained permanent damage to one eye. (*East Fort Myers, Florida*)

## Tuesday, March 23, 1965

John T. King was driving in Bangor, Maine when he came upon an enormous domed disk hovering above the road in front of him. As King approached the object, his car experienced electrical difficulties and his lights dimmed and his radio went dead. King, believing he was in mortal danger, pulled out a Magnum pistol he had in the car and began to shoot at the UFO out the driver's side window of his vehicle. It isn't certain whether any of the shots actually hit the UFO, but after King fired the third time, the craft took off at what he later described as "tremendous speed." (*Bangor, Maine*)

## Wednesday, October 17, 1973

Falkville, Alabama Police Chief Jeff Greenhaw received a call at his home at ten o'clock in the evening from a woman who claimed she had seen a UFO land in field outside of town. Greenhaw drove to the area and came upon a silver-suited figure with an antenna on its head standing in the middle of the road. Greenhaw got out of his car and took four photos of the entity with a Polaroid camera he had brought with him. When Greenhaw got back in the cruiser and turned on the blue lights on top of his car, the figure started to run away. Greenhaw pursued the entity in his police cruiser and yet was not able to catch him, even at speeds as high as 35–40 miles per hour. Greenhaw ended up ditching the car on the side of the road and the entity vanished into the night. Greenhaw's photos showed a humanoid figure dressed in some kind of silver foil-like suit and it is impossible to determine if it is actually an alien or someone pulling off a hoax. Following the incident, Greenhaw's house was burned down and he was fired from his job. (*Falkville, Alabama*)

## Wednesday, August 13, 1975

US Air Force Staff Sergeant Charles L. Moody saw a disk-shaped UFO descend towards him as he sat in his car one night on the outskirts of Alamogordo, New Mexico watching for meteors. As the UFO neared, his car would not start and he heard what he described as a high-pitched whining sound emanating from the UFO. He then remembered seeing shadowy figures inside the UFO, feeling numb, and then watching as the UFO flew away. Driving home, he realized that it was one-and-half hours later than when he first saw the UFO, and that he could not account for the missing time. The following day his back was inflamed, there was a puncture wound above his spinal column, and a few days later, he developed a head-to-toe body rash. Months later Moody recalled being abducted and remembered being in telepathic communication with aliens who were approximately 4 foot, 8 inches tall, had light gray skin, large heads and eyes, and slit-like mouths. Moody also remembered lying on a table inside the UFO and having a rod inserted into his back. (*Alamogordo, New Mexico*)

## Monday, November 17, 1986

While approaching Alaska's Anchorage Airport at 39,000 feet for a stopover on a flight from France to Japan, Japan Air Lines Flight 628 pilot Captain Kenju Terauchi saw an object twice the size of an aircraft carrier and rimmed with lights fly alongside his plane for over 30 minutes. The object was tracked on radar by Anchorage Airport's air traffic controllers. Captain Terauchi later commented on his "Close Encounter of the First Kind." He told the media, "We were carrying Beaujolais from France to Japan. Maybe [the aliens] wanted to drink it." (*Anchorage, Alaska*)

## Monday, October 17, 1988

The thirty-five-year-old manager of a jewelry store (who wishes to remain anonymous), now a writer, was headed to work, traveling east on US Route 1 in East Haven, Connecticut. There were very few clouds in the sky and the weather was clear and sunny. At 9:45 a.m., the man stopped for a red light and just happened to look up into the northeastern sky. There he saw clearly an oval-shaped silver object moving rapidly in a perfectly straight line in a northerly direction. At first the man thought he was seeing a high-flying jet, since East Haven was near a major airport and seeing (and hearing) all kinds of aircraft in the area at all hours of the day and night was not an uncommon experience. But then the flying object suddenly stopped in mid-air, reversed course, traveled in the opposite direction for a few seconds, and then disappeared. It is not known if others witnessed this UFO and no terrestrial explanation for the sighting has been determined to this day. (*East Haven, Connecticut*)

## Friday, July 18, 1997

Sandra Brown and John Gardner reported seeing a daylight disk hovering above Highway 35 in Sparta, Georgia. The UFO remained in one place for approximately five minutes and was observed by Brown and Gardner the entire time. Brown described the UFO to the *Gallatin County News* as "round, silver, had no lights and was a little bigger than a car." She also told the newspaper, "It kind of just sat there. It hung in the sky in one place and then, in the blink of an eye, it disappeared. I know people will think we're crazy, but we really saw this." There has been no terrestrial explanation for the UFO since the sighting. (*Sparta, Kentucky*)

## Saturday, April 25, 1998

At 11:50 p.m. a college astronomy major named Kevin observed a red light "zigzagging at high speed" across the sky approximately 40 degrees above the southern horizon in Comer, Georgia. Kevin's brother also witnessed the UFO and the two men later used 7x35 binoculars to get a closer look at the lights, which they reported remained horizontal relative to the surface of the Earth the entire time they observed it. The brothers continued watching the UFO until it disappeared at 12:30 a.m. on Sunday, April 26, 1998. The UFO's final sighted position was approximately 20 to 30 degrees above the southern horizon. Of his sighting, Kevin later said, "I know the difference between a planet, heavenly body, and a flying object." (*Comer, Georgia*)

# QUOTES

## Amazing & Credible Quotations about UFOs

*"There's UFOs over New York . . ."*

John Lennon

Who says high-ranking people have not gone on record about the existence of UFOs? Check out this collection of quotations about UFOs by people ranging from World War 1 flying ace Eddie Rickenbacker to President Harry S Truman. Let's put it this way: these guys oughta know, right?

**From an August 1954 statement by Royal Air Force Commanding Officer Air Chief Marshall Lord Hugh Dowding:**

"Of course the flying saucers are real—and they are interplanetary . . . the cumulative evidence for the existence of UFOs is quite overwhelming and I accept the fact of their existence."

**From a statement by World War I flying ace Captain Eddie Rickenbacker:**

"Flying saucers are real. Too many good men have seen them, that don't have hallucinations."

**From a 1952 FBI memorandum by Air Intelligence Officer Commander Boyd:**

"The objects sighted may possibly be from another planet . . . at the present time there is nothing to substantiate this . . . but the possibility is not being overlooked. . . . Intense research is being carried out by Air Intelligence. . . . The Air Force is attempting in each instance to send up jet interceptor planes."

**From a December 1952 memorandum by Assistant Director of Scientific Intelligence H. Marshall Chadwell to the Director of the CIA:**

"Sightings of unexplained objects at great altitude and traveling at high speeds in the vicinity of major US defense installations are of such nature that they are not attributable to natural phenomena or known types of aerial vehicles."

**From a 1957 statement by Darling Observatory astronomer Dr. Frank Halstead:**

"Many professional astronomers are convinced that [flying] saucers are interplanetary machines."

**From a statement by Japan's Air Self-Defense Force's Chief of Air Staff, General Kanshi Ishikawa:**

"UFOs are real and they may come from outer space. . . . UFO photographs and various materials show scientifically that there are more advanced people piloting the saucers and motherships."

**From a statement by US General Douglas MacArthur:**

"The nations of the world will have to unite; for the next war will be an interplanetary war. The nations of the Earth must some day make a common front against attack by people from other planets."

**From a statement by CIA director and NICAP board member Vice Admiral Roscoe H. Hillenkoetter:**

"Through official secrecy and ridicule, many citizens are led to believe the unknown flying saucers are nonsense."

**A notation added to a memo by longtime FBI Director J. Edgar Hoover:**

"I would [aid the Army Air Force in its investigation] but before agreeing to it we must insist upon full access to the disks recovered. For instance in the LA case the Army grabbed it and would not let us have it for cursory examination."

**From a statement by Director-General of the International Air Transport Association, Knut Hammarskjöld:**

"UFOs [are] probably observation machines from outer space."

**From a statement issued on July 8, 1947, by Roswell Army Air Base Public Information Officer Lieutenant Walter Haut:**

"The many rumors regarding the flying disk became a reality yesterday when the intelligence office of the 509th Bomb Group of the 8th Air Force Roswell Army airfield was fortunate enough to gain possession of a disk through the cooperation of one of the local ranchers and the Sheriff's office of Chaves County. The flying object landed on a ranch near Roswell sometime last week."

**From an April 4, 1950, statement by US President Harry S. Truman:**

"I can assure you that flying saucers, given that they exist, are not constructed by any power on earth."

**From a statement by former Air Force intelligence officer and US military UFO investigator Steve Lewis:**

"That movie *Close Encounters of the Third Kind* is more realistic than you believe."

**From an early 1990s interview with former Soviet President Mikhail Gorbachev:**

"The phenomenon of UFOs does exist, and it must be treated seriously."

**A comment Ronald Reagan reportedly made to Steven Spielberg during a screening of *ET: The Extraterrestrial* at the White House in 1982. The remark was recounted to *Alien Contact* author Timothy Good by Michael Luckman, who apparently overheard it firsthand.**

"There are probably only six people in this room who know how true this is."

**NASA X-15 pilot Major Robert White, after a July 17, 1962, UFO sighting during a test flight:**

"There are things out there! There absolutely is!"

***Apollo 17* Commander Eugene Cernan, quoted in a 1973 Los Angeles Times article:**

"I've been asked [about UFOs] and I've said publicly I thought they were somebody else, some other civilization."

**From a 1979 statement by former chief of NASA Communications Systems Maurice Chatelaine:**

"All *Apollo* and *Gemini* flights were followed, both at a distance and sometimes also quite closely, by space vehicles of extraterrestrial origin—flying saucers, or UFOs, if you want to call them by that name. Every time it occurred, the astronauts informed Mission Control, who then ordered absolute silence."

**From a statement by NASA Deputy Public Relations Director Albert M. Chop:**

"We are being watched by beings from outer space."

# The Shapes of Things to Come: The 10 Most Commonly Reported UFO Types

## The 10 Most Commonly Reported UFO Types

The National Investigations Committee on Aerial Phenomena (NICAP) was founded in 1956 by US Navy physicist Thomas Townsend Brown. From the mid-fifties until his removal in 1969, US Marine Corps Major Donald E. Keyhoe helmed the Committee. During his tenure, Keyhoe and his field researchers compiled the most comprehensive listing of reported UFO physical types to date. These ten UFO types were chronicled in NICAP's 1964 publication, *The UFO Evidence.*

NICAP was dissolved in 1973, and its papers and materials were turned over to Dr. J. Allen Hynek.

Here is a look at the ten most commonly reported UFO types, as determined by NICAP over almost a decade of research and investigation.

1. **Flat Disk:** These are either round or oval, have flat bottoms, or are lens-shaped. (This is the reason why lenticular clouds have often been reported erroneously as UFOs.) "Coin"-shaped flying disks have also been reported in this category.

2. **Domed Disk:** These are described as round and looking like a hat or a helmet. They always have a domed top.

3. **Saturn Disk:** These disks resemble the planet Saturn and share the characteristic of being "double-domes," with a rounded top and bottom with a thicker ring protruding from the center of its body. Variations of this disk include an elliptical or "winged oval" disk; a diamond-shaped disk (a deeper bottom dome than top); and the most commonly reported Saturn-shaped.

4. **Hemispherical Disk:** These disks are often reported in three variations: the "Parachute," with a high, domed top; the "Mushroom," with a domed top and extended bottom; and the "Half Moon," which has a domed top and flattened bottom.

5. **Flattened Sphere:** These disks are spherical with flattened and elongated tops and bottoms. Occasionally they are seen to have a point or a peak on the top of the upper "deck."

6. **Spherical:** These are essentially airborne globes that are sometimes metallic looking and that sometimes appear as a ball of glowing light. They almost always appear perfectly round from every angle.

7. **Elliptical:** These disks look like footballs or eggs and are sometimes seen to have two distinct levels with a non-protruding center "belt." (Unlike the Saturn disks that have a pronounced protruding ring around their middle.)

8. **Triangular:** These craft look like flying triangles and sometimes are described as pear-shaped or tear-drop-shaped. Cubes, tetrahedrals, and crescents also fall into this category

9. **Cylindrical:** Cylindrical UFOs resemble rockets or cigars and sometimes are reported to have windows or portholes running along their length.

10. **Light Source Only:** These UFOs resemble planets or stars and in many cases that is exactly what they are.

## FACT

### CUFOS' 4 Types of UFO Evidence

These are the four types of scientific evidence the J. Allen Hynek Center believes are the most important when evaluating a UFO sighting.

**1. PHYSICAL TRACES:**

"Compressed and dehydrated vegetation, broken tree branches, and imprints in the ground have all been reported. Sometimes a soil sample taken from an area where a UFO had been seen close to the ground will be determined, through laboratory analysis, to have undergone heating or other chemical changes not true of the control sample."

**2. MEDICAL RECORDS:**

"Medical verification of burns, eye inflammation, temporary blindness, and other physiological effects attributed to encounters with UFOs—even the healing of previous conditions—can also constitute evidence, especially when no other cause for the effect can be determined by the medical examiner."

**3. RADARSCOPE PHOTOS:**

"A tape of traces from a radar screen on which a 'blip' of a UFO is appearing is a powerful adjunct to a visual sighting, because it can be studied at leisure instead of during the heat of the moment of the actual sighting."

**4. PHOTOGRAPHS:**

"While it might seem that photographs would be the best evidence for UFOs, this has not been the case. Hoaxes can be exposed very easily. But even those photos that pass the test of instrumented analysis and/or computer enhancement often show nothing more than an object of unknown nature, usually some distance from the camera, and very often out of focus. For proper analysis of a photo, the negative must be available and the photographer, witnesses, and circumstances must be known. In a few exceptional cases, photos do exist that have been thoroughly examined and appear to show a structured craft."

# *Popular Mechanics* Magazine's Most Credible UFO Sightings

*Popular Mechanics offers no opinion on whether these mysterious flying machines originate from secret military airstrips here on Earth or spaceports somewhere "out there." We do, however, feel comfortable making one prediction: When the shell of security surrounding UFOs finally cracks, it will be because one of the sightings we present here provided the wedge.*

—from the Introduction to
"Six Unexplainable Encounters"
by Jim Wilson

The cover of the July 1988 issue of *Popular Mechanics* shouted "UFOs: The 6 Sightings They Can't Explain Away." The story inside by Jim Wilson was titled "Six Unexplainable Encounters" and looked at precisely that: six UFO reports that are still a mystery and which do not (so far) seem to have a terrestrial explanation.

This list takes a look at the magazine's choices and provides additional information about these "unexplainable" sightings.

**McMinnville, Oregon (Thursday, May 11, 1950)**

Paul Trent, a farmer *Life* magazine described in its June 26, 1950 issue as "an honest individual," took two black-and-white photographs of a UFO hovering above his farm on Thursday, May 11, 1950.

Edward Condon, head of the University of Colorado UFO project (usually referred to as the Condon Committee) wrote the following about the McMinnville sighting: "This is one of the few UFO reports in which all factors investigated, geometric, psychological, and physical, appear to be consistent with the assertion that an extraordinary flying object, silvery, metallic, disk-shaped, tens of meters in diameter, and evidently artificial, flew within sight of two witnesses. It cannot be said that the evidence positively rules out a fabrication, although there are some physical factors such as the accuracy of certain photometric measures of the original negatives which argue against a fabrication."

Around 7:30 p.m. on the evening of May 11, 1950, Mrs. Paul Trent saw a flying disk-shaped object moving slowly towards her, west-southwest across the sky above her and her husband's farm. Mrs. Trent called to her husband Paul, who was in the house. Paul came out, saw the object, and ran to his car to fetch his camera. The camera was actually in the house, however, and Mrs. Trent ran inside and returned with it. Paul Trent took the camera from his wife and immediately snapped one photo. He quickly rewound the film, and then snapped another photo. Mr. and Mrs. Trent then watched as the object turned toward the west and quickly flew out of sight.

The UFO the Trents saw and photographed was a flat-bottomed disk with a pointed top. Later, Trent described the craft as being about the size of a "good-sized parachute canopy without the strings." He described its color as "silvery-bright mixed with bronze."

The chain of possession of the Trent's extraordinary photos is circuitous indeed. After the pictures were developed in 1950, they were displayed in the window of a local bank where Frank Wortmann, a friend of the Trents, worked. On June 8, the pictures were published in the *McMinnville*

*Telephone Register*, a local newspaper. The *Register* story was subsequently picked up by the International News Service and reprinted all over the world. The appearance of the story prompted *Life* magazine to borrow the negatives from Bill Powell, the *Register* reporter who had borrowed them from the Trents to write the story. *Life* published the pictures in their June 26, 1950, issue and then claimed to have misplaced the negatives. The pictures were found 17 years later in the files of United Press International, which had bought out INS. They then went to the Condon Committee and then back to UPI. They were ultimately found in the *Register*'s files five years later by UFO investigator Bruce Maccabee, who said, "In retrospect it is probably a good thing that the negatives were 'lost' between 1950 and 1967 because they were reasonably well protected during this time, and therefore the photographic information was only minimally degraded."

Over the past four decades, there has been considerable photographic analysis of the Trent photos. UFO debunker Philip Klass asserts that the shadows seen on the Trents' garage prove that the photos could only have been taken in the morning, not the evening as claimed by the Trents. Later analysis, however, disproves this theory and most UFO researchers agree that the Trents were incapable (both morally and intellectually) of constructing an elaborate UFO hoax—buttressed by the fact that, to this day, the Trents have not received a cent for their pictures or their story.

Conclusion? Fabrication or hoax, hallucination, military traffic, and natural explanation have all been ruled out. As of now, the McMinnville sighting can only be explained as an authentic sighting of an extraterrestrial craft. Adding to the mystery is the fact that the McMinnville photographs are almost identical to a UFO photograph taken by a pilot in Rouen, France in the summer of 1954.

### Trindade, Brazil (Thursday, January 16, 1958)

The Saturn-shaped UFO seen by forty-seven members of a Brazilian oceanographic and meteorological research team and photographed by a civilian photographer is the kind of sighting that drives UFO debunkers nuts.

The credibility of the observers is usually paramount when determining the authenticity—or the completely bogus nature—of a UFO sighting or photograph. The witnesses to the Trindade sighting were forty-seven crew members of the naval vessel *Almirante Saldanha*, many of whom were professional oceanographers or meteorologists—men for whom science is their forte and accurate observation their code of honor.

On the evening of January 16, 1958, the entire crew of the research vessel observed fast-flying disks in the skies above the island of Trindade in the south Atlantic. Photographer Almiro Baurana snapped six photographs, and after they were developed and analyzed, it was determined that Baurana had captured an object with a diameter of 50 feet, traveling at a speed of approximately 600 miles per hour. "It glittered at certain moments," Baurana later told the Brazilian magazine *O Cruzeiro*, "perhaps reflecting the sunlight, perhaps changing its own light—I don't know. It was coming over the sea, moving toward the point called the Galo Crest. I had lost 30 seconds looking for the object, but the camera was already in my hands, ready, when I sighted it clearly silhouetted against the clouds. I shot two photos before it disappeared behind Desejado Peak. . . . The object remained out of sight for a few seconds—behind the peak—reappearing bigger in size and flying in the opposite direction, but lower and closer then before, and moving at a higher speed." Baurana ultimately shot six photos, of which two were useless because of being jostled while aiming his camera.

In order to assure himself that there would be no tampering with the photos or negatives, the captain of the research vessel insisted that Baurana develop the pictures in a makeshift darkroom onboard ship, wearing only his swimming trunks. The pictures were later examined by a Brazilian photoanalysis firm and judged to be authentic.

The pictures show a flattened and elongated globe with a ring surrounding it (much like the rings of Saturn) moving across the sky above a mountainous ridge.

UFO debunker Donald H. Menzel initially said that the photos showed a plane flying through fog. He later recanted that conclusion and said that the pictures had been faked by Baurana. Both of Menzel's opinions were invalidated by a report released in 1978 by Ground Saucer Watch (GSW), a highly regarded organization that specializes in analysis of UFO photographs. Their report read in part:

"The UFO image is over 50 feet in diameter. The UFO image in each case reveals a vast difference from the photographer/camera. The photographs show no signs of hoax (i.e., a hand-thrown or suspended model). The UFO image is reflecting light and passed all computer tests for an image with substance. The image represents no known type of aircraft or experimental balloon. Digital densitometry reveals a metallic reflection. . . . We are of the unanimous opinion that the Brazilian photos are authentic and represent an extraordinary flying object of unknown origin."

Also of great significance is the report issued by NICAP regarding the Brazilian photos:

"Weighing all the facts, we conclude that the pictures appear to be authentic. They definitely are one of the potentially most significant series of UFO photographs on record, so that clarification of the incident and additional analysis is strongly desirable."

### Hillsdale, Michigan (Monday, March 21, 1966)

This is the UFO sighting that transformed J. Allen Hynek from a skeptic to a believer, and it is a doozy.

On the evening of March 21, 1966, a female student at Hillsdale College reported to the police a strange glowing object visible in the sky above the women's dormitory. County Civil Defense director William Van Horn responded to the report and confirmed that a glowing object was indeed maneuvering above a swamp near the school. Other witnesses that evening included the dean of the college and reports of between 40 and 140 students.

Dr. Hynek (then working as a consultant for Project Blue Book) was instructed by the Air Force to investigate this sighting and several others that had taken place in the area in the days prior to the March 21 sighting. Hynek complied, and during a hastily convened press conference on Friday, March 25, Hynek said that a possible cause for the sightings could have been burning swamp gas.

The media (conveniently ignoring the word "possible" in his statement) latched onto Hynek's desperate explanation, and he was immediately ridiculed for allegedly trying to "explain away" these amazing sightings of what some had seen clearly as *objects* as nothing but the release of swamp gas.

Dr. Albert Hibbs of the California Institute of Technology later told the media, "The characteristics of swamp gas do not accord with what was reported."

A chagrined Hynek changed his position two weeks later when a soil analysis report was issued. The analysis had been done on soil from the area of the sighting and the report stated that on the spot where the swamp gas had "touched down," the soil contained higher than normal levels of radiation.

The report also determined that the soil was contaminated with boron—an element not normally found in soil and which is used to retard nuclear chain reactions.

The Air Force later issued a statement regarding the Hillsdale sightings suggesting that "certain young men have played pranks with flares"; that photographs taken show "trails made as a result of time exposure of the rising crescent moon and the planet Venus"; that the observers who reported seeing an object were "no closer than 500 yards—a distance which does not allow details to be determined"; and that "this unusual and puzzling display" was due to an exceptionally mild winter and "the particular weather conditions of that night."

Notwithstanding the Air Force's "swamp gas" explanation (which conveniently does not address the increased levels of radiation and boron found at the site), there has yet to be a credible explanation for the Michigan sightings; especially the Hillsdale sighting that made the esteemed Dr. J. Allen Hynek change his mind.

### Zanesville, Ohio (Sunday, November 13, 1966)

This sighting by Ralph Ditter, a barber and amateur astronomer, resulted in—*if* they are authentic—two of the most astonishing photographs of UFOs ever taken.

The photos were taken with a Polaroid camera and show a hat-shaped UFO hovering above Ditter's house in broad daylight. The pictures are remarkably clear and quite detailed, and the objects seen in the photos resemble other craft seen elsewhere around the world. The craft in the Ditter photos shows a remarkable resemblance to the craft photographed on a Santa Ana, California highway in 1965 by Rex Heflin.

Skeptics, however, point to the incredible clarity and detail in the Zanesville photos as evidence of fakery. And yet similar craft had been seen in the area months earlier, with reports of their sighting submitted by extremely credible law enforcement officials.

The Ditter photos still stand as quintessential examples of visual evidence of UFOs: they are, unquestionably, *Unidentified* Flying Objects. Whether or not they show actual extraterrestrial craft has yet to be determined with certainty.

### Sheraz, Iran (Sunday, October 8, 1978)

Iran has had many UFO sightings over the past several decades, and many have involved unexplained evasive airborne craft that shows up on radar and then either eludes scrambled fighters or just zips away.

This Sheraz sighting was somewhat different since the craft that was seen and photographed was almost identical to a UFO photographed four months earlier in Iran by a sixteen-year-old student up late studying for exams.

The first photo of the mysterious craft was taken in June 1978 by Jamshid Saiadipour. The craft he photographed resembled a craft that Iranian pilots had reported seeing in the air near the Teheran airport earlier that year.

Then, on Sunday, October 8, Franklin Youri took a photo of a UFO hovering above his home in western Iran that looked exactly like the one Saiadipour had seen, as well as the one Iranian pilots had reported seeing.

Skeptics believe that the two photos show a secret stealth jet known as Tacit Blue, and the craft seen do resemble the top-secret jet. And yet the photo taken by Youri seems to show a craft hovering above the roof of his house at a very low altitude, a feat most jets—even top secret ones—are probably not capable of achieving.

These Iranian sightings may turn out to be photos of the Tacit Blue—and yet there are still too many unanswered questions to state with certainty that that is actually what they are.

### Bentwaters, England (Saturday, December 27, 1980)

This sighting in 1980 is important because it took place at a location where one of the most astonishing radar/visual reports of a UFO took place in 1956.

On August 13, 1956, RAF radar picked up crafts moving at speeds of up to 9,000 m.p.h. at Bentwaters/Lakenheath in Suffolk, England. The objects were later seen visually, but the photos taken didn't come out.

Even the relentlessly skeptical *Condon Report* didn't know what to make of this 1956 sighting, stating, "In summary, this is the most puzzling and unusual case in the radar-visual files. The apparently rational, intelligent behavior of the UFO suggests a mechanical device of unknown origin as the most probable explanation of this sighting."

Unusual lights were again observed fourteen years later in the same location, Rendlesham Forest, just outside the Bentwaters US Air Force Base. After two days of seeing these lights, security personnel entered the forest with floodlights, Geiger counters, radios, and other gear and all present observed a 20-foot wide, 30-foot tall triangular-shaped craft hover and then take off. The object may have had some kind of mechanical legs.

The following day, three circular depressions were found in the woods where the craft had been sighted. The depressions were 7 inches in diameter and approximately 1½ inches deep. Radioactivity readings taken from this site showed radiation readings twenty-five times the normal range. There were also broken tree limbs in the surrounding area.

UFO experts believe that an extraterrestrial craft landed on the Bentwaters woods and left physical evidence. The increased radiation levels are similar to the findings from the 1966 Hillsdale, Michigan sightings.

So far, there has been no terrestrial explanation for this occurrence.

# UFO Sightings by Astronauts

*During several space missions, NASA astronauts have reported phenomena not immediately explainable; however, in every instance NASA determined that the observations could not be termed "abnormal" in the space environment.*

—from NASA's official February 14, 1997,
Statement on the US Government's
Involvement in Investigating UFO Reports

Oh, yeah?

Read through these accounts of astronaut UFO sightings and decide for yourself.

**MAJOR GORDON COOPER (*Mercury* astronaut)**

Astronaut Gordon Cooper had three notable UFO sightings. The first was in 1951 while flying an F-86 Saberjet over Germany. At that time he reported the sighting and chasing a group of metallic, disk-shaped objects that were capable of aerial maneuvers no earthly craft could then achieve.

Cooper's second sighting was on May 2, 1957, at Edwards Air Force Base. He was project manager for the Flight Test Center at Edwards and later, in an interview with Lee Siegel, said that a flying disk had landed at the Base. Cooper said it hovered above the ground "and then it slowly came down and sat on the lake bed for a few minutes." Cooper also said that the entire incident was filmed and the cameramen "all agreed that [the craft] was at least the size of a vehicle that would carry normal-size people in it."

Cooper also is reported as saying, "It was a typical circular-shaped UFO. Not too many people saw it because it took off at quite a sharp angle and just climbed out of sight."

In 1993, Cooper told UFOlogist Michael Hesemann the following: "I had a crew that was filming an installation of a precision landing system we were installing out on the dry lake bed, and they were there with stills and movies and filmed the whole installation and they came running in to tell me that this UFO, a little saucer, had come right down over them, put down three gear, and landed about 50 yards from them, and as they proceeded to go on over to get a closer shot of it, it lifted up, put the gear in, and disappeared in a rapid rate of speed . . . I had to look up the regulations on who I was to call to report this, which I did, and they ordered me to immediately have the film developed, put it in a pouch, and send them by the commanding general's plane to Washington, which I did. And that was the last I've ever heard of the film."

Cooper had very strong feelings about what he had seen: "I think it was definitely a UFO. However, where it came from and who was in it is hard to determine, because it didn't stay around long enough to discuss the matter!"

Cooper's third sighting was during a May 1963 *Mercury* space mission. During his final orbit, Cooper told the Muchea, Australia tracking station that he was watching a glowing, greenish-colored object that quickly approached his capsule. The object was confirmed by Muchea's radar to be a solid object and reported on TV by NBC. After Cooper landed, however, the media was forbidden to ask him about the green UFO he had seen during his space flight.

Cooper believed that many UFOs were of extraterrestrial origin. In 1978, a letter from Cooper was read before the United Nations General Assembly. In the letter, Cooper stated bluntly, "I believe that these extra-terrestrial vehicles and their crews are visiting this planet from other planets, which obviously are a little more technically advanced than we are here on earth." He also expressed his feeling that "we need to have a top level coordinated program to scientifically collect and analyze data from all over the earth concerning any type of encounter, and to determine how best to interface with these visitors in a friendly fashion."

### JAMES McDIVITT (*Gemini 4* astronaut)

While orbiting earth during the *Gemini 4* mission in June 1965, Astronaut James McDivitt reported seeing a cylindrical-shaped object with protuberances resembling antenna-like extensions. According to the *Condon Report*, the "appearance was something like the second phase of a Titan . . . [but] it was not possible to estimate its distance."

McDivitt said it had a white or silver appearance and that it definitely had angular extension—it did not appear simply as a point of light against a dark sky, but rather as a defined object with visible dimensionality.

One still photo was taken of the object, as well as some black-and-white film footage. McDivitt and his fellow astronaut Ed White both agreed that at the time of the sighting, they felt it might be necessary to take evasive action to avoid a collision.

Upon landing, the film was turned over to NASA, and ultimately, three or four pictures from the film footage were released. McDivitt was adamant, however, that the pictures NASA released were *not* of the object he had seen.

McDivitt later gave his opinion that what he had seen was probably an unmanned satellite, although research by the Condon Committee determined that at the time of the sighting, there was no unmanned satellite in the area that would satisfy both the physical description given by McDivitt and the distance estimates. "The suggestion . . . that this was a satellite," Dr. Condon wrote, "has not been confirmed . . . by a definite indication of a known satellite."

Condon also revealed that during a conversation with McDivitt, the astronaut stated that during another space flight, he had once seen what he described as "a light" moving across the sky relative to the star background (which was immobile). McDivitt told Condon that he could not make out any details regarding this sighting.

## JAMES LOVELL AND FRANK BORMAN (*Gemini 7* astronauts)

During their December 1965 *Gemini 7* flight, Lovell and Borman saw something strange, and the following exchange took place between the spacecraft and NASA's Houston Control Center [emphasis added]:

SPACECRAFT: *Gemini 7* here. Houston, how do you read?
HOUSTON: Loud and clear, 7, go ahead.
SPACECRAFT: *Bogey* at 10 o'clock high.
HOUSTON: This is Houston. Say again 7.
SPACECRAFT: Said *we have a bogey* at 10 o'clock high.
HOUSTON: Roger. *Gemini 7*, is that the booster *or is that an actual sighting?*
SPACECRAFT: *We have several,* looks like debris up here. *Actual sighting.*
HOUSTON: You have any more information? Estimate distance or size?
SPACECRAFT: We *also* have the booster in sight.
HOUSTON: Understand you *also* have the booster in sight, Roger.
SPACECRAFT: Yeah, we have a very, very many—look like hundreds of little particles banked on the left out about 3 to 7 miles.
HOUSTON: Understand you have many small particles going by on the left. At what distance?
SPACECRAFT: Oh about—it looks like *a path of the vehicle* at 90 degrees.
HOUSTON: Roger, understand that they are about 3 to 4 miles away.

SPACECRAFT: They are passed now, they are in polar orbit.
HOUSTON: Roger, understand that they were about 3 or 4 miles away.
SPACECRAFT: That's what it appeared like. That's Roger.
HOUSTON: Were these particles in addition to the booster and *the bogey* at 10 o'clock high?
SPACECRAFT: Roger . . . I [Lovell] have the booster on *my* side, it's a brilliant body in the sun, against a black background with trillions of particles on it.
HOUSTON: Roger. What direction is it from you?
SPACECRAFT: It's about at my [Lovell's] 2 o'clock position.
HOUSTON: Does that mean that it's ahead of you?
SPACECRAFT: It's ahead of us at 2 o'clock, slowly tumbling.

Based on the above conversation, the *Condon Report* came to the following conclusion [emphasis again added]:

> *"The general reconstruction of the sighting . . . is that in addition to the booster traveling in an orbit similar to that of the spacecraft there was another bright object (bogey) together with many illuminated particles. It might be conjectured that the bogey and particles were fragments from the launching of* Gemini 7, *but this is impossible if they were traveling in a polar orbit as they appeared to the astronauts to be doing."*

**NEIL ARMSTRONG, EDWIN "BUZZ" ALDRIN, AND MICHAEL COLLINS (*Apollo 11* astronauts)**

In 1986, Neil Armstrong told UFOlogist Timothy Good that he did not see anything abnormal during his *Apollo* space flights and that "All observations on all *Apollo* flights were fully reported to the public." However, for years, people have pointed to published transcripts from the *Apollo 11* technical debriefing in which Armstrong, Aldrin, and Collins all discussed a sighting of a cylindrical UFO while on their way to the Moon. Here are selected excerpts from the debriefing:

ALDRIN: The first unusual thing that we saw . . . was one day out or pretty close to the moon. It had a sizable dimension to it . . . we called the ground and were told the [Saturn IV booster] was 6,000 miles away.

ARMSTRONG: [The object] was really two rings. Two connected rings.

COLLINS: It was a hollow cylinder. But then you could change the focus on the sextant and it would be replaced by this open-book shape. It was really weird.

Also, the following conversation transcript has been circulating among UFO buffs for some time now. Allegedly, during a NASA symposium, a university professor had the following exchange with *Apollo 11* astronaut Neil Armstrong. The professor reportedly insisted on remaining anonymous, and this conversation has not been confirmed by Neil Armstrong (and considering his comment to Timothy Good cited above, either this exchange is a hoax, or Armstrong was more revealing to this mysterious professor than he had been to researchers.) It is presented here for its curiosity value and as an example of the kind of material that circulates in the UFO "underground."

PROFESSOR: What *really* happened out there with *Apollo 11*?

ARMSTRONG: It was incredible. Of course we had always known there was a possibility, the fact is, we were warned off! [by the Aliens]. There was never any question then of a space station or a moon city.

PROFESSOR: How do you mean "warned off"?

ARMSTRONG: I can't go into details, except to say that their ships were far superior to ours both in size and technology. Boy, were they big! And menacing! No, there is no question of a space station.

PROFESSOR: But NASA had other missions after *Apollo 11*?

ARMSTRONG: Naturally . . . NASA was committed at that time, and couldn't risk panic on Earth. But it really was a quick scoop and back again.

Adding fuel to the extraterrestrial fire is the persistent rumor that ham radio operators who were able to pick up NASA broadcasts during the *Apollo 11* mission recorded Neil Armstrong excitedly telling Mission Control that he was seeing huge alien spacecraft lined up on the moon watching him and his fellow astronauts as they gamboled on the lunar terrain. In 1979, Maurice Chatelain, the former chief of NASA Communications Systems, confirmed in an interview that Neil Armstrong had reported seeing two UFOs on the rim of a lunar crater.

### DONALD SLAYTON (*Mercury* astronaut)

In 1951, years before he became an astronaut, Slayton saw a UFO over Minneapolis while testing a P-51 fighter. In a published interview, Slayton described his close encounter: "I was at about 10,000 feet on a nice, bright, sunny afternoon. I thought the object was a kite, then I realized that no kite is gonna fly that high. As I got closer it looked like a weather balloon, gray and about three feet in diameter. But as soon as I got behind the darn thing it didn't look like a balloon anymore. It looked like a saucer, a disk. About the same time, I realized that it was suddenly going away from me—and there I was, running at about 300 miles per hour. I tracked it for a little way, and then all of a sudden the damn thing just took off. It pulled about a 45-degree climbing turn and accelerated and just flat disappeared."

# FACT

### Jimmy Carter's Answers to the 37 Questions on His 1973 NICAP UFO Sighting Report

*I am convinced that UFOs exist because I've seen one . . . It was a very peculiar aberration, but about 20 people saw it . . . It was the darndest thing I've ever seen. It was big; it was very bright; it changed colors. and it was about the size of the moon. We watched it for 10 minutes, but none of us could figure out what it was . . .*

—Jimmy Carter, *The National Enquirer*, June 8, 1976

**1. Name:**

*Jimmy Carter*

**2. Date of Observation and Time:**

*October 1969; 7:15 P.M. EST*

**3. Locality of Observation:**

*Leary, Georgia*

**4. How long did you see the object?**

*10–12 minutes.*

5. Please describe weather conditions and the type of sky; i.e., bright daylight, nighttime, dusk, etc.

    *Shortly after dark.*

6. Position of the Sun or Moon in relation to the object and to you:

    *Not in sight.*

7. If seen at night, twilight, or dawn, were the stars or moon visible?

    *Stars.*

8. Were there more than one object?

    *No.*

9. Please describe the object(s) in detail. For instance, did it (they) appear solid, or only as a source of light; was it revolving, etc.? Please use additional sheets of paper, if necessary.

    *[Not answered.]*

10. Was the object(s) brighter than the background of the sky?

    *Yes.*

11. If so, compare the brightness with the sun, moon, headlights, etc.

    *At one time, as bright as the moon.*

12. Did the object(s)—

A. Appear to stand still at any time? *Yes*

B. Suddenly speed up and rush away at any time? --

C. Break up into parts or explode? --

D. Give off smoke? --

E. Leave any visible trail? --

F. Drop anything? --

G. Change brightness? *Yes*

H. Change shape? *Size*

I. Change color *Yes*

*Seemed to move toward us from a distance, stopped—moved partially away—returned, then departed. Bluish at first, then reddish, luminous, not solid.*

13. Did object(s) at any time pass in front of, or behind of, anything? If so, please elaborate giving distance, size, etc., if possible.

*No.*

14. Was there any wind? If so, please give direction and speed.

*No.*

15. **Did you observe the object(s) through an optical instrument or other aid, windshield, windowpane, storm window, screening, etc.? What?**

*No.*

16. **Did the object(s) have any sound? What kind? How loud?**

*No.*

17. **Please tell if the object(s) was (were)—**

A. Fuzzy or blurred. --
B. Like a bright star. --
C. Sharply outlined. *X*

18. **Was the object—**

Self-luminous? *X*
Dull finish? --
Reflecting? --
Transparent? --

19. **Did the object(s) rise or fall while in motion?**

*Came close, moved away—came close then moved away.*

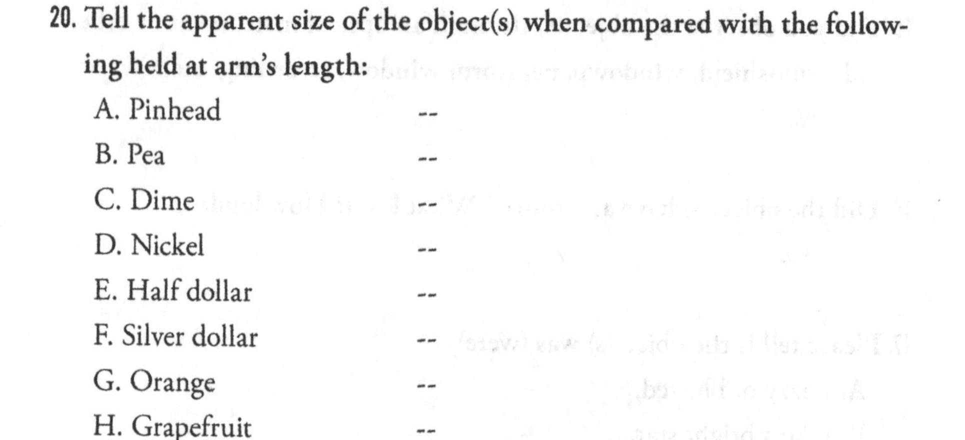

**20.** **Tell the apparent size of the object(s) when compared with the following held at arm's length:**

A. Pinhead --

B. Pea --

C. Dime --

D. Nickel --

E. Half dollar --

F. Silver dollar --

G. Orange --

H. Grapefruit --

I. Larger --

Or, if easier, give apparent size in inches on a ruler held at arm's length.

*About the same as moon, maybe a little smaller. Varied from brighter/ larger than planet to apparent size of moon.*

**21.** **How did you happen to notice the object(s)?**

*10–12 men all watched it. Brightness attracted us.*

**22.** **Where were you and what were you doing at the time?**

*Outdoors waiting for a meeting to begin at 7:30 P.M.*

**23.** **How did the object(s) disappear from view?**

*Moved to distance then disappeared.*

**24. Compare the speed of the object(s) with a piston or jet aircraft at the same apparent altitude.**

*Not pertinent.*

**25. Were there any conventional aircraft in the location at the time or immediately afterwards? If so, please elaborate.**

*No.*

**26. Please estimate the distance of the object(s).**

*Difficult. Maybe 300–1000 yards.*

**27. What was the elevation of the object(s) in the sky?**

*About 30° above horizon.*

**28. Names and addresses of other witnesses, if any.**

*Ten members of Leary Georgia Lions Club.*

**29. What do you think you saw?**

A. Extraterrestrial device? *[Not answered.]*

B. UFO? *[Not answered.]*

C. Planet or star? *[Not answered.]*

D. Aircraft? *[Not answered.]*

E. Satellite? *[Not answered.]*

F. Hoax? *[Not answered.]*

G. Other (please specify). *[Not answered.]*

30. Please describe your feelings and reactions during the sighting. Were you calm, nervous, frightened, apprehensive, awed, etc.? If you wish your answer to this question to remain confidential, please indicate with a check mark. (use a separate sheet if necessary.)

*[Not answered.]*

31. Please draw a map of the locality of the observation showing north; your position; the direction from which the object(s) appeared and disappeared from view; the direction of its course over the area; roads, towns, villages, railroads, and other landmarks within a mile.

*Appeared from West—about 30° up.*

32. Is there an airport, military, governmental, or research installation in the area?

*No.*

33. Have you seen other objects of an unidentified nature? If so, please describe these observations, using a separate sheet of paper.

*No.*

34. Please enclose photographs, motion pictures, news clippings, notes of radio or television programs (include time, station and date, if possible) regarding this or similar observations, or any other background material. We will return the material to you if requested.

*None.*

35. Were you interrogated by Air Force investigators? By any other federal, state, county, or local officials? If so, please state the name and rank or title of the agent, his office, and details as to where and when the questioning took place

*[Not answered.]*

36. We should like permission to quote your name in connection with this report. This action will encourage other responsible citizens to report similar observations to NICAP. However, if you prefer, we will keep your name confidential.

*You may use my name.*

37. Date of filling out this report.

*9-18-73*

# Irritating Posterity: Dr. J. Allen Hynek's Responses to 7 Misconceptions about UFO Sighting Reports

*I am reminded of a conversation I once had with U Thant, the late Secretary General of the United Nations, during my days as a skeptic. We had been discussing UFOs and interstellar travel, and he asked me whether I thought extraterrestrials might possibly visit our world. I responded that as an astronomer I found the distances and the times necessary to make the journey so great as to preclude it entirely. U Thant looked at me, arched his eyebrows, and said: 'You know, I am a Buddhist, and we believe in life elsewhere.' I told him that as an astronomer I did too, but that the physical conditions, especially the length of time involved in journeys from outer space, seemed insuperable. The Secretary General paused, leaned back in his chair, and said, 'Ah, but what may seem like years to us, may be just a day or two to others.' And so it may.*

—J. Allen Hynek, from the Prologue
to *The Hynek UFO Report*

In 1997, Barnes & Noble Books, in association with the Estate of J. Allen Hynek, published a new edition of Hynek's classic 1977 work, *The Hynek UFO Report.* This new edition boasted a foreword by acclaimed UFO authority Jacques Vallee and was a welcome addition to the library of authoritative books about the UFO phenomenon and what Hynek obviously believed was the US government's official debunking policy regarding sightings.

What this book made clear was that by 1977, Hynek had converted from skeptic to curious agnostic when it came to the existence of alien-controlled UFOs and extraterrestrial visitations to Earth. *The Hynek UFO Report* was a comprehensive review of the most mysterious and unexplainable Project Blue Book reports, accompanied by Hynek's astute and open-minded interpretation of these sightings. [See the chapter "The 218 Unexplained *Project Blue Book* UFO Sightings Made by Military Personnel" for more on a very specific subset of the Blue Book archives.]

But ten years earlier, Hynek was already rejecting the standard debunking answers provided by the United States government when it came to UFO reports and was beginning to speak out publicly about such duplicitousness and deception. Hynek knew a cover-up when he saw one and in the rare, January 1967 issue of *Fate* magazine, he published an "Open Letter" to the editors of *Science Magazine* in Washington, D. C., chastising them for what seemed to him to be a deliberate and neglectful shunning of legitimate UFO news. "I cannot dismiss the UFO phenomenon with a shrug," he wrote. "I have begun to feel that there is a tendency in twentieth-century science to forget that there will be a twenty-first-century science, and indeed, a thirtieth-century science, from which vantage points our knowledge of the universe may appear quite different than it does to us." Hynek concludes that "We suffer, perhaps from temporal provincialism, a form of arrogance that has always irritated posterity."

This list looks at Hynek's 1967 responses to seven accepted misconceptions about UFO sightings and reports—and the people who make these reports. [Hynek's description of each of the misconceptions is verbatim; his explication is paraphrased.]

1. **"Only UFO 'buffs' report UFOs."**: Hynek makes the point that the exact *opposite* of this misconception is closer to the truth. The true believers don't need to file UFO reports or acquire

and disseminate evidence of alien crafts and visitations. They are utterly convinced of the truth of the existence of UFOs. The result of this is that most reports actually come from "regular people"; people, Hynek says, "who have *not* given much or any thought to UFOs, generally considering them 'bunk' until shaken by their own experience."

2. **"UFOs are reported by unreliable, unstable, and uneducated people."**: Hynek admits that this statement is true in some cases, but notes that *most* UFOs are reported by "reliable, stable, and educated people"—people with a higher-than-average level of intelligence. "Dullards," he writes, "rarely overcome the inertia inherent in getting down to making a written report."

3. **"UFOs are never reported by scientifically trained people."**: "This is unequivocally false," Hynek states with certainty. He writes that scientifically savvy people frequently make UFO reports, but they usually add the disclaimer that they "don't believe in flying saucers." They may not believe in them, but they all admit to seeing something beyond their understanding and having no logical explanation and, as scientists are trained to do, they write their experiences up for the record.

4. **"UFOs are never seen clearly at close range, but are seen under conditions of great uncertainty and always reported vaguely."**: Hynek writes that there are *countless* reports that are the exact opposite of this definition and that *these* are the ones he feels should be examined and investigated by qualified scientific authorities and agencies. Hynek writes that he is *not* talking about (or encouraging investigation of) the sightings and reports that correspond with Statement No. 4.

5. **"The Air Force has no evidence that UFOs are extraterrestrial or represent advanced technology of any kind.":** Hynek states that this is a true statement—but only because *truly unidentified* sighting reports simply *cannot* be used to respond to this statement. "As long as there are 'unidentifieds,'" he writes, "the question must obviously remain open."

6. **"UFO reports are generated by publicity.":** Hynek makes the point that the reason for this apparent "me too" reaction to sighting reports is that many people come forward *only* after they hear of other people having similar experiences. Many people are hesitant to speak about a sighting or file a report for fear of being perceived as a wacko, and yet, when there are several *publicized* reports of sightings, these "silent sighters" do venture forth and speak out about what they, too, saw. "One cannot deny that there is stimulated emission of UFO reports," he concludes, "but it is unwarranted to assert that this is the sole cause of high incidence of UFO reports."

7. **"UFOs have never been sighted on radar or photographed by meteor or satellite tracking cameras.":** Not true, Hynek writes, since he personally saw what he described as "oddities" on satellite tracking photographs and radar reports—anomalies that he admits could have very well been balloon trails or unusual aircraft "but they never have been positively identified." And that is the crux of the UFO paradox: Aberrant airborne phenomena that has not been identified does *not* mean that they are unidentifiable, Hynek explains. And that is why he believes that serious scientists should be taking a closer look at the truly unidentified sightings.

Hynek concludes his letter to *Science* magazine by reaffirming that his "concern stems not from hearing reports selected for their sensational aspects, but from noting a pattern emerge after many years of 'monitoring' the phenomenon. This pattern [repeated similar flight behavior and physical characteristics] suggests that 'something is going on.'"

## Mysterious NASA Scientist

Here's a little-known story from a former NASA scientist, who shall remain nameless. After the crash at Roswell, one of the still-living ETs was transported to Wright Field in Dayton, Ohio, where a select group of scientists and military began testing it to understand how it ticked. But this was a sentient ET, who knew that the tests being conducted upon it were killing it. In spring, 1948, the class at the National Air War College were secretly detailed to Wright Field, where they viewed the Roswell debris. Then they were allowed to view the ET, who communicated to them that it was being killed by those running biological tests on it. This ET was kept by the Eisenhower administration until it died. One of the people allowed to view this top secret life form was J. Allen Hynek. However, in exchange for the access to this ET, Hynek had to work for Blue Book, as the official debunker of UFO claims until the Air Force shut down the operation. Afterwards, Hynek was free to advance his own theories, helping research the Hudson Valley Sightings and working with Steven Spielberg on his *Close Encounters of the Third Kind*. Hynek never disclosed that he had seen the ET, but spent the rest of his life advancing theories supporting the reality of UFO phenomena.

## The J. Allen Hynek Center for UFO Studies' 3 Kinds of "Relatively Distant" UFO Sightings

These definitions are from the official J. Allen Hynek Center for UFO Studies (CUFOS) brochure that is distributed to people interested in learning more about the organization.

They are, for all intents and purposes, the defining criteria for distant UFO sightings.

1. **Nocturnal Lights:** "These are sightings of well-defined lights in the night sky whose appearance and/or motion are not explainable in terms of conventional light sources. The lights appear most often as red, blue, orange or white. They form the largest group of UFO reports."

2. **Daylight Disks:** "Daytime sightings are generally of oval or disk-shaped, metallic-appearing objects. They can appear high in the sky or close to the ground, and they are often reported to hover. They can seem to disappear with astounding speed."

3. **Radar-Visual Cases:** "Of special significance are unidentified 'blips' on radar screens that coincide with and confirm simultaneously visual sightings by the same or other witnesses. These cases are infrequently reported."

# Possible Explanations for Daylight UFO Sightings

In his 1979 book, *The UFO Handbook: A Guide to Investigating, Evaluating and Reporting UFO Sightings*, Allen Hendry, the chief investigator and managing editor of the *International UFO Reporter*, presents a sober-minded look at UFO sightings, with an eye toward separating the truly *unidentified* flying objects from the IFOs—the *identified* flying objects.

Hendry interviewed 1,300 witnesses for his book and was ultimately able to distinguish most of the explainable sightings from the unexplainable ones.

Hendry's hard-nosed, pragmatic approach to UFO sightings—and his deliberate weeding out of the sightings that are most assuredly *not* extraterrestrial in origin—makes the remaining sightings covered in his book (which boasts an Introduction by J. Allen Hynek) all the more impressive.

As part of his research, Hendry compiled the thirteen most common explanations for daylight UFO sightings, and this list takes a look at his important findings.

### Weather balloons

Weather balloons are the most commonly reported "UFOs." They range in size from large research balloons to small weather balloons and the wide range of shapes, colors, and the way the sun reflects off them make them easily misidentified as unknown flying craft.

### Aircraft

Airborne airplanes are often difficult to identify with certainty, and many people have reported ordinary winged aircraft as UFOs.

### Meteors

Meteors are rarely seen during the daylight hours, but when they are, they can often be mistaken for strange flying craft. They usually streak across the sky (sometimes covering several states) and thus are visible to a great many people, some of whom will be certain that they have just seen an extraterrestrial spacecraft.

### Venus

A few times a year Venus is bright enough to be seen in the sky during the day. Clouds, ground haze, and normal, involuntary movements of the human eye can all make Venus look like she is moving across the sky.

### Lenticular clouds

These large, circular, domed clouds can look amazingly like mysterious flying objects when seen from a distance. They are rare, but they last a long time and thus, are frequently photographed when they do appear.

### Birds

Flocks of birds can look like oddly-behaving aircraft, especially when a flock darts around the sky and is seen from a distance.

### Visual disorders

People with eye problems often see things that are not there!

### Kites and kite wire

When seen from a distance, kites can occasionally appear as strange aircraft or flying objects.

### Sundogs

When the sun reflects off ice crystals in the air, sometimes the reflection appears as an airborne glowing object.

### Experimental aircraft

People will often report as a UFO a flying craft they have never seen before—even though it may just be a working prototype for a new plane or helicopter.

### Model aircraft

Radio-controlled small planes, especially when they appear out of nowhere, can often be interpreted as a UFO.

### Windborne objects

Have you ever thrown a Frisbee or a hat into the air and snapped a picture of it? Windborne objects can easily appear as a UFO in a photograph.

### Mirages

Pure hallucination. And yet sometimes, people report mirages as UFO sightings.

# FACT

## The 5 Times Ronald Reagan Delivered His "Alien Invasion" Speech

At a Geneva Summit in 1985, Ronald Reagan gave a historic speech in which he discussed the possibility of an alien invasion of Earth. The most important passage was the following:

> *How easy his task and mine might be in these meetings that we held if suddenly there were a threat from some other species from another planet outside in the Universe. We would forget all the little local differences that we have between our countries, and we would find out once and all that we really are human beings here on this Earth together.*

Reagan gave this speech a total of five times in the next three years, as listed below.

1. At a Geneva Summit in 1985.
2. At the United Nations in 1987.
3. At the Washington Summit in 1987.
4. At the National Strategic Forum in 1988.
5. At the Moscow Summit in 1988.

Why did the President of the United States—the most powerful and influential leader on the planet—talk about an alien invasion of Earth, not once, not twice, but an astonishing *five times*?

You may speculate on the answer to that important question and come to your own conclusions.

Among the conclusions you may consider is what did Ronald Reagan know, when did he know it, and how did he know it? One story involves a briefing by Caspar Weinberger, the former Secretary of Defense, who attended a briefing meeting with President Reagan during which the entire history of UFOs, the military, and national security was raised. President Reagan was astounded by the amount of information his advisors disclosed. Thus, when he had the chance to meet with Gorbachev in Iceland, the president suggested that the two leaders should not be at odds when they faced a common enemy, a statement he also made to the United Nations General Assembly. For a little bit of substantiation of this story, consider an interview that filmmaker Steven Spielberg gave years later in which he told the story of President Reagan talking to him after the screening of his film *ET*. According to Spielberg, in the White House screening room filled with guests, President Reagan said to him in a loud voice that everyone heard, "Only you and I know that this film is not fiction." The others in the room chuckled, but Spielberg said that President Reagan was not laughing, he was truly serious. This is why President Reagan might have really been in the loop of UFO disclosure.

# "A High Degree of Strangeness": The 5 Kinds of Close Encounters

## The 5 Kinds of Close Encounters

*Middle-American strangers become involved in the attempts of benevolent aliens to contact earthlings.*

—The Videohound's Golden Movie Retriever's description of Steven Spielberg's 1977 movie, *Close Encounters of the Third Kind*

In 1972, astronomer and former Project Blue Book researcher Dr. J. Allen Hynek, the founder of CUFOS, created a ranking system for recording human encounters with extraterrestrial craft and extraterrestrials.

Hynek's classification system was so useful and so specific, it is now used as the standard tool for the accurate chronicling of UFO and alien sightings and reports.

Here is a look at Hynek's original three rankings and the two "spin-off" rankings that have come into common usage in the past several years.

1. **Close Encounters of the First Kind (CE-I cases):** UFO sighting reports in which very bright lights or objects are seen by witnesses at distances of 600 feet or less. CUFOS states, "Though the

witness observes a UFO nearby, there appears to be no interaction with either the witness or the environment."

In *The Hynek UFO Report*, Dr. Hynek wrote that a CE-I is "often a frightening experience, and always an awesome one, but when it is over there are no visible marks or evidence of it. The event is so unusual, so traumatic generally, that even when a camera has been available . . . we have no record of its having being used."

2. **Close Encounters of the Second Kind (CE-II cases):** CUFOS states, "These encounters include details of interaction between the UFO and the environment which may vary from interference with car ignition systems and electronic gear to imprints or burns on the ground and physical effects on plants, animals and humans."

In *The Hynek UFO Report*, Dr. Hynek wrote that CE-II cases "seem to describe a UFO that is capable of leaving physical traces on its surroundings, but whose behavior does not correspond with our present technology and certainly not with the technology of 1947 or 1948. Since we are always inclined to believe in physical evidence—what we can see with our own eyes—CE-IIs tend to offer more convincing evidence than any other type of UFO sighting that the UFO phenomenon is 'real.'"

3. **Close Encounters of the Third Kind (CE-III cases):** CUFOS states, "In this category, occupants of a UFO—entities that are humanlike ('humanoid') or not humanlike in appearance—have been reported. There is usually no direct contact or communication with the witness. However, in recent years, reports of incidents involving very close contact—even detainment of witnesses—have increased."

In his book, Dr. Hynek wrote that reports of Close Encounters of the Third Kind, "whether they be single- or multiple-witness cases, are characterized by a high degree of strangeness and by the complete bewilderment of the witnesses."

CE-III cases are unique because of (in most cases) the undeniable credibility of the witnesses. People who have seen alien beings are loathe to come forward, and they almost always shun publicity. They are fearful of being labeled insane and are also terrified of being abducted (and becoming a CE-IV statistic!) due to the fact that they have now actually *seen* the aliens and must, therefore, be silenced.

Alien sighting reports are *convincing*: The people who see extraterrestrials were usually not out on the bog with binoculars or a telescope searching the skies for alien spaceships. In many cases, these percipients were driving to work, washing their car, or just sitting around taking it easy. As Dr. Hynek has said, CE-IIIs seem to be "a most real phenomenon of undetermined origin."

4. **Close Encounters of the Fourth Kind (CE-IV cases):** These cases consist of a close encounter with an alien being and/or an abduction by an extraterrestrial. Dr. Hynek lumped abductions with CE-III cases, but lately, this category has been used to describe specific cases in which humans interact with aliens—beyond simply seeing them on the ground or in their craft—or cases in which humans are kidnapped by aliens.

Abduction cases are ubiquitous in the UFO literature, and two excellent books on the subject are John Mack's seminal 1994 work, *Abduction: Human Encounters with Aliens*, and Jenny Randles' 1994 book, *Alien Contacts and Abductions: The Real Story from the Other Side.* Both are highly recommended for those interested in this terrifying (and possibly earth-shattering) facet of UFOlogy.

5. **Close Encounters of the Fifth Kind (CE-V cases):** This is a relatively new designation for Close Encounter cases and is generally used to describe mutual, intentional communication with an extraterrestrial, such as cases in which humans channel aliens; governments work with aliens in a mutually reciprocal way (as the US Government is alleged to be doing with the Grays); or certain individuals serve as spokespersons for specific extraterrestrial species (e.g., Billy Meier and the Pleiadians).

Some UFOlogists have also used this designation to describe visitations by unknown beings during which no spacecraft can be seen. This application, however, seems to blur the line between astral manifestations such as poltergeists, spiritual manifestations such as angels, waking dreams and hallucinations, and the genuine experience of an alien presence.

# 8 Characteristics of "The Oz Factor"

*We're not in Kansas anymore.*

—Dorothy, *The Wizard of Oz*

"The Oz Factor" is a term coined by renowned UFOlogist Jenny Randles to describe an altered state of consciousness experienced by UFO percipients during a Close Encounter. Randles defines it as a "sensation of being isolated, or transported from the real world into a different environmental framework." She states that this feeling is "one of great importance to our understanding of UFOs," and it manifests very specific—and recognizable—attributes.

Here is a look at a few of the common characteristics of Close Encounters that Randles calls the "Oz Factor."

1. **Timelessness:** Many UFO percipients report "missing time" experiences in which minutes and sometimes hours pass, seemingly in seconds. I interviewed a man who had a four-hour missing time experience in which he actually drove miles along a highway with no recollection of his actions.

2. **Being transported to another place:** People often report feeling as though they had somehow left the real world they inhabit before the sighting, and had been transported to another

dimension of reality, even if nearby sights and sounds remain the same.

3. **Absolute silence:** A characteristic of the Oz Factor in UFO experiences is the sudden and complete lack of all extemporaneous sound—including traffic noise, birds chirping, and the normal sounds of city life and activity. All vanish, and the environment becomes strangely quiet (and "strange" is a common adjective used to describe this eerie silence).

4. **Motionlessness:** Some UFO percipients report feeling a stillness in the immediate area of the sighting and occasionally feeling as though they cannot move.

5. **Bright lights:** Some UFO percipients report strange blue lights (and sometimes bright white lights) during experiences and sightings.

6. **Odd sounds:** Snapping noises, odd humming sounds, buzzing, and other weird, intermittent sounds have been reported during UFO sightings.

7. **Feelings of apprehension:** Many abductees report fear and apprehension at the beginning of an abduction or, occasionally, even during a sighting.

8. **A feeling of sensory deprivation:** Drowsiness, physical weakness, and lightheadedness are often reported during sightings and as part of an abductee's last memory before being abducted.

# FACT

### CUFOS' 4 Things to Do If You See a UFO

The Hynek Center is serious about chronicling UFO sightings, and the organization makes a concerted effort to educate the public about what to do if they experience a UFO sighting. These four "Things To Do" are part of their official brochure and serve as an important set of guidelines if the day ever comes when you look up into the sky and see something you suspect may not be of this planet.

1. "First, try to get another witness—as many other witnesses as possible."

2. "If you have a camera handy, take as many photographs as possible. Don't worry about getting a perfect picture. Get as much background or foreground detail into the photo as possible."

3. "Immediately after your sighting, make complete notes of everything you can remember. Write down the appearance, color, motion and size of the UFO, as well as what you were thinking and feeling when you had the experience. Write down the names and addresses and phone numbers of other witnesses."

4. "If the UFO touched the ground, do what you can to protect the area—but don't disturb the area. Take photographs of the area to document it."

At the conclusion of this set of instructions is the following:

> Most importantly, *report your UFO sighting.*
> Call the Center for UFO Studies any day of the week, any time of the day or night, directly at (773) 271-3611.
> An investigator will contact you as soon as possible.

# Celebrity UFO Sightings

This chapter looks at some of the high-profile actors, musicians, sports stars, and politicians who have gone public with their UFO sightings and other types of close encounters. These stories bring to mind something I once heard on some TV show you might have seen: The truth is out there.

Indeed.

### Muhammad Ali

The champ admits to having seen at least sixteen UFOs during his life. One was "cigar-shaped" and appeared over the New Jersey Turnpike. Another looked like "a huge electric light bulb"; he saw it in Central Park. "If you look into the sky in the early morning you see them playing tag between the stars," he said.

### David Bowie

David "*The Man Who Fell To Earth*" Bowie claimed to have seen many UFOs when he was growing up in England. In fact, he saw UFOs so frequently that he published a magazine about his sightings. "They came over so regularly," he once said in an interview, "we could time them." Bowie also reported, "Sometimes they stood still, other times they moved so fast it was hard to keep a steady eye on them." This kind of explains "Major Tom" and the whole *Ziggy Stardust* thing, don't you think?

### Jimmy Carter

Former President Carter claims to have seen a UFO on January 6, 1969. He was governor of Georgia at the time; Carter said it changed colors and was as big as the moon. There is conjecture today that what he (and the others he was with) saw was actually the planet Venus. Although his description of the object certainly doesn't sound like a stationary planet. [See "Jimmy Carter's Answers to the 37 Questions on his 1973 NICAP UFO Sighting Report" in this volume.]

### Jamie Farr

Farr played Corporal Klinger on the hit TV series M*A*S*H, and claims to have once seen an erratically flying light in Yuma, Arizona. He says it stopped in midair above his car. No further information about this sighting ever came to light.

### Jackie Gleason

The Great One was extremely interested in the paranormal and the occult and claimed to have seen several UFOs in London and Florida (as well as having seen retrieved alien craft and corpses). He possessed a massive library of books about UFOs and the supernatural and subscribed to many UFO journals. Gleason even named his home in Peekskill, New York, the Mothership.

### Dick Gregory

According to activist Gregory, he saw three red and green lights dance in the sky for close to an hour one night.

### John Lennon

The former Beatle told his UFO story to biographer Ray Coleman during an interview in Lennon's Dakota apartment. "Look, it's true," he told Coleman. "I was standing, naked, by this window leading on to that roof when an oval-shaped object started flying from left to right. After about twenty minutes it disappeared over the East River and behind the United Nations Building. I wonder if it had been carrying out some research here." John's assistant and one-time lover May Pang also saw the UFO and later said, "It looked like a flattened cone with a brilliant light on top." "They all think I'm potty," John continued, "but it's true. I shouted after it, 'Wait for me, wait for me!'" John also told Coleman that he immediately called the police to report the sighting and was told that there had been other reports of flying saucers seen over New York that evening. "But I didn't tell them who was on the phone," he went on. "I didn't want newspaper headlines saying, 'Beatle Sees Flying Saucer.' I've got enough trouble with the immigration people already." Coleman then asked John if he had been smoking pot or drinking prior to seeing the UFO. "No," he replied. "God's honest truth. I only do that on weekends or when I see Harry Nillson."

### Gordon MacRae

Six months before he died, during an appearance on a TV talk show, actor Gordon MacRae (*Oklahoma, Carousel*) stated that he had been a security sergeant at Wright-Patterson Air Force Base in July 1947. He then told the fantastic story of being ordered to stand watch over a large pallet covered with a tarpaulin that had been brought in under tight security. MacRae remembered being specifically ordered not to remove the tarp under any circumstances but admitted that curiosity got the better of him and his fellow guards. They did, in fact, peek under the tarp MacRae said that they all saw four small humanoid creatures laid out on the pallet. MacRae also said that he believed the crashed saucer from Roswell, New Mexico was being stored at Wright-Patterson at this time.

### Richard Nixon

Reportedly, President Nixon was fully aware of the crash and retrieval of alien spacecraft and alien bodies and was known to have visited the repository of these extraterrestrial artifacts at Homestead Air Force Base in Florida. Nixon even supposedly once arranged a 1973 tour of the facility (complete with a viewing of the alien bodies) for his good friend Jackie Gleason, who, according to his wife, returned home from the experience quite shaken. UFOlogist Timothy Good filed a Freedom of Information Act request to Homestead Air Force Base for all files and records pertaining to Gleason's visit (and the alien repository) but was told no such records existed. Gleason never went on record confirming the story.

### Ronald Reagan

In 1974, when Ronald Reagan was governor of California, he, his pilot, and two security agents all saw a UFO while flying aboard a Cessna Citation near Bakersfield, California. It was between nine and ten o'clock at night and Reagan's pilot, Bill Paynter, described the object as "a big light flying a bit behind [the] plane." He also reported that "It seemed to be several hundred yards away," and that it was "a fairly steady light until it began to accelerate, then it appeared to elongate." Paynter also said the object then "went up at a 45-degree angle", a behavior reported in many other UFO sightings. Ronald Reagan himself later described the incident to a reporter for the *Wall Street Journal*, saying, "We followed it for several minutes. All of a sudden, to our utter amazement it went straight up into the heavens. When I got off the plane, I told Nancy about it. And we read up on the long history of UFOs . . . " Norman Miller, the *Wall Street Journal* reporter interviewing Reagan, then took the leap and asked Reagan point blank if he believed in UFOs. Reagan immediately distanced himself from the UFO believers (even though it was obvious that he believed what he saw qualified as a UFO), telling Miller, "Let's just say that on the subject of UFOs I'm an agnostic."

### William Shatner

You'd almost think he was joking but, no, Shatner—known around the galaxy as Captain James T. Kirk of the Starship *Enterprise*—was serious when he reported that a silver spacecraft flew over him in the Mojave Desert as he pushed his inoperative motorcycle. Shatner also claims to have received a telepathic message from the beings in the craft advising him which direction to walk. (The least they could have done was give him a jump-start, no?) Shatner later told UFOlogist Timothy Green Beckley that a "ghostly motorcyclist" had led him to safety, seemingly confusing a ghost with alien visitors. (So which was it, Captain?) In any case, Shatner says he saw an "object glistening in the heavens" when he finally made his way back to town, obviously believing that this "object" was the UFO from which he had received his either extraterrestrial or metaphysical guidance.

# PART III
# A UFO WHO'S WHO

# 103 Important People in UFOlogy

Here is a look at 102 of the most important people in the study of UFOs and the extraterrestrial presence on Earth.

This "Who's Who" is meant to serve as an introduction to these influential people, not as a compendium of comprehensive biographies. We refer you to the sources section for books about many of these individuals that go into far more detail than we can provide here.

1. **Adamski, George:** One of the earliest (if not the first) modern era contactees. In 1953 Adamski published *Flying Saucers Have Landed*, his story of lifelong UFO sightings, travels on alien spacecraft, and interactions and encounters with aliens. Adamski still has many supporters, although some of his stories are now obviously impossible—such as walking on the surface of Venus. Also, some of the photos he claimed to have taken of alien spacecraft have been determined to be fakes. Nevertheless, Adamski's followers believe most of what he reported and his name holds a hallowed place in the annals of UFOlogy.

2. **Andreasson, Betty:** Famous abductee about whom there are three books: *The Andreasson Affair*, *The Andreasson Affair, Phase Two*, and *The Watchers*. Andreasson claimed that

on January 25, 1967, aliens entered her home through the walls. She was then abducted by an alien named Quazgaa, medically examined, and she heard the "Voice of God." The investigation into the Andreasson affair is ongoing to this day.

3. **Andrus, Walter H., Jr.:** The current International Director of the world's largest UFO organization, the Sequin, Texas-based Mutual UFO Network, most commonly known as MUFON.

4. **Armstrong, Lieutenant:** An Air Force pilot who saw six flying disks four days after Kenneth Arnold's encounter in 1947.

5. **Arnold, Kenneth:** The pilot whose June 24, 1947, sighting of nine UFOs resulted in the coining of the term "flying saucers" and the beginning of the modern age of UFO sightings. Arnold witnessed the nine objects flying in formation at speeds between 1,300 and 1,700 m.p.h. He said their formation was approximately five miles long and that each object was approximately two-thirds the size of a DC-4. Media attention to Arnold's sighting was immediate and rabid, and soon reports of UFOs were flooding military and local authorities. On July 12, 1947, Arnold issued the following statement: "I observed these objects not only through the glass of my airplane but turned my airplane sideways where I could open my window and observe them with a completely unobstructed view (without sunglasses)." Arnold's sighting is still considered a milestone in the annals of UFOlogy fifty years later.

6. **Atkinson, Colonel Ivan C.:** The Air Force officer who initiated the creation of the Condon Committee.

7. **Bailey, Al and Betty:** Husband and wife who claimed to have witnessed an Adamski meeting with aliens.

8. **Barry, Bob:** Religious fundamentalist and head of a UFO organization who believes that Steven Spielberg's movie *Close Encounters of the Third Kind* was part of a government propaganda program to condition people on Earth to accept alien contact without fear or panic.

9. **Bethurum, Truman:** Contactee and author of *Aboard a Flying Saucer.*

10. **Blanchard, Colonel William H.:** The Base Commander of the Roswell Army Air Base in 1947.

11. **Bonilla, José A. Y.:** An 1883 astronomer who took one of the first UFO photos.

12. **Bowen, Charles:** One of the first editors of the respected British magazine *Flying Saucer Review.*

13. **Boyd, Commander:** The Air Intelligence officer who conceded in a 1952 FBI memo that UFOs may be from another planet. [See Boyd's memo in this volume.]

14. **Brace, Lee Fore:** British Merchant Marine who saw two UFOs in the Persian Gulf in May 1880.

15. **Brazel, William "Mac":** The manager of the ranch on which the Roswell crash occurred.

16. **Brown, Thomas Townsend:** The founder of NICAP.

17. **Burns, Governor Haydon:** The Governor of Florida in April 1966 who was buzzed by a UFO while aboard his campaign aircraft and who later went public with the account of his encounter.

18. **Bush, Dr. Vannevar:** The Head of Majestic 12.

19. **Birnes, William:** Author and UFOlogist; wrote *The Day After Roswell*; co-author of *The Big Book of UFO Facts, Figures & Truth*; starred in the TV series *UFO Files, Ancient Aliens,* and *UFO Hunters*. His motto is, "If UFOs weren't real, because of their prevalence throughout world history and in our own skies today, our governments would have to invent them just to explain away the phenomena."

    This notion originally came from a CIA agent talking about the Stealth fighter, who said, "If UFOs weren't real, we'd have to invent them to explain away the weapons we have."

20. **Carpenter, Scott:** A US astronaut alleged to have photographed UFOs while in space in 1962 but who later denied it.

21. **Carter, President Jimmy:** The only American President to publicly admit to having seen a UFO in October 1969. [See Carter's Report in this volume.]

22. **Chadwell, H. Marshall:** The US Assistant Director of Scientific Intelligence who acknowledged in a December 1952 memo to the Director of the CIA that UFOs might be extraterrestrial. [See Chadwell's memo in this volume.]

23. **Clark, Jerome:** Renowned UFOlogist and the author of the seminal three-volume *UFO Encyclopedia*. [See the interview with Jerome Clark in this volume.]

24. **Condon, Dr. Edward:** The head of the debunking Condon Committee. Dr. Edward Condon was Dean Edward Condon of the University of Colorado at Boulder, who was asked by the head of the Air Force to write an introduction to the Colorado Report on UFOs to dismiss them as not a real phenomenon and not a national security threat. He did this to clear his name from charges of subversion leveled against him by prosecutor Roy Cohn, who claimed that Condon's theories of quantum physics were subversive because they were based upon European scientists. Writing the introduction to the Colorado report in which he debunked UFO sightings put him back in the government's good graces for helping out the Air Force shut down Blue Book.

25. **Cooper, Gordon:** A US astronaut who admitted a UFO sighting.

26. **Corbin, Lieutenant-Colonel Lou:** Former Army Intelligence officer who repeatedly released information about UFOs and government UFO investigations.

27. **Crick, Francis:** The molecular biologist who publicly espoused the theory that extraterrestrials had purposely "infected" the Earth. Crick referred to his theory of ETs' fertilizing Earth and other planets as "panspermia."

28. **Creighton, Gordon:** The current editor of *Flying Saucer Review*.

29. **Däniken, Erich Von:** Popularizer of the "Ancient Astronaut" theory and author of the *Chariots of the Gods* books.

30. **Davis, Kathie [alias]:** Multiple abductee who was the subject of Budd Hopkins' book *Intruders*.

31. **Donovan, Major General William:** The Head of the Office of Strategic Services who concluded that "foo fighters" were natural phenomena and not a German secret weapon or extraterrestrial in origin.

32. **Dowding, Air Chief Marshal Lord High:** A Royal Air Force Commanding Officer who, in an August 1954 statement, averred that it is a certainty that UFOs are real. [See Dowding's statement in this volume.]

33. **Doyle, Sir Arthur Conan:** "Sherlock Holmes" author who was a spiritualist and who reported a paralyzing bedroom visitation by an unknown entity.

34. **DuBose, Colonel Thomas Jefferson:** The Air Force aide who confirmed that the material recovered in the Roswell crash was quickly shipped to Wright Field to buttress the fabricated "weather balloon" story and satisfy press interest.

35. **Edwards, Frank:** UFO expert, radio commentator, journalist, and author of *Flying Saucers–Serious Business* and *Flying Saucers–Here and Now!*

36. **Eisenhower, President Dwight D.:** US President who was rumored to have visited Edwards Air Force Base and viewed crashed alien spacecraft and recovered alien bodies. The White House stated at the time that Eisenhower had gone to the dentist.

37. **ET:** The alien star of Steven Spielberg's blockbuster film *ET–The Extraterrestrial.*

38. **Ezekiel:** The biblical prophet who recorded what many interpret as a UFO/alien encounter. [See "40 Biblical Passages That Might Describe UFOs or Extraterrestrial Contacts" in this volume.]

39. **Fish, Marjorie:** Ohio Teacher who, in 1969, used Betty Hill's original star drawing to construct a three-dimensional star map that pinpointed the origin of the aliens who kidnapped the Hills as the stars Zeta Reticuli 1 and 2.

40. **Ford, President Gerald:** US President whose efforts, as a Congressman, led to the establishment of the Condon Committee.

41. **Forrestal, Secretary James V.:** US Secretary of Defense and Secretary of the Navy who was rumored to have been part of the Majestic 12 panel. Forrestal committed suicide in 1949.

42. **Fort, Charles:** Legendary writer and researcher into unexplained phenomena. Fort's name became an adjective—paranormal phenomena are now often referred to as "Fortean events"—and his work spawned a magazine, *Fortean Times*.

43. **Friedman, Stanton T.:** Nuclear Physicist and author (*Crash at Corona*), who was the original investigator of the Roswell incident and who is now a respected and much sought-after authority on UFOs and the alien presence.

44. **Fry, Daniel W.:** Contactee and author of *The White Sands Incident*.

45. **Gaia:** Earth as a living entity. Some UFO researchers believe that UFOs may be a physical manifestation of Gaia's consciousness.

46. **Gardner, Norma:** A clerk at Wright Patterson Air Force Base in the 1950s who, on her deathbed, revealed details about cataloging UFO sightings, parts of alien spacecraft, autopsy reports on humanoid bodies, and other facts.

47. **Gilligan, Governor John:** Governor of Ohio who, in October 1973, saw a UFO and admitted it.

48. **Glass, Lieutenant Henry F.:** Pilot and the first UFO sighter to report (in August 1946) a craft with rows of windows.

49. **Godfrey, Arthur:** Nationally known entertainer and pilot who had a UFO encounter in June 1965 and reported it.

50. **Goethe, Johann Wolfgang Von:** The author of *Faust*. Goethe reported an alien entity encounter in his autobiography in 1768.

51. **Greenhaw, Police Chief Jeff:** Police Chief who had a UFO sighting and an alien entity encounter in Falkville, Alabama in October 1973.

52. **Halstead, Dr. Frank:** A renowned astronomer who stated that many professional astronomers believe that UFOs are interplanetary machines. [See Halstead's statement in this volume.]

53. **Halt, Lieutenant-Colonel Charles:** A high-ranking USAF officer who witnessed UFOs in Rendlesham Forest in England in 1980.

54. **Hammarskjöld, Knut:** The Director-General of the International Air Transport Association who stated that, based on the reports he had received, he believed that UFOs were from outer space. [See Hammarskjöld's statement in this volume.]

55. **Haut, Lieutenant Walter:** The Public Information Officer for the Roswell Army Air Base who issued the statement that the US Air Force had recovered a crashed flying saucer.

56. **Heaton, Peter Henniker:** A staff writer for the *Christian Science Monitor* who wrote a scathing article blasting the Condon Report's blatant anti-UFO bias.

57. **Henderson, Major Paul W.:** The Pease Air Force Base officer who confirmed that the UFO that Betty and Barney Hill claimed had abducted them had been detected by radar at Pease Air Force Base.

58. **Hill, Betty and Barney:** Two of the most important abductees in UFOlogy. The Hills were abducted in September 1961 and two years later consented to hypnotic regression, during which they recounted their abduction. Betty Hill drew a star map which turned out to be accurate. In 1966, John G. Fuller wrote a book about the Hills' experience called *An Interrupted Journey.*

59. **Hillenkoetter, Vice Admiral Roscoe H.:** The first director of the CIA, and a board member of the NICAP. Hillenkoetter admitted that citizens are being misled about UFOs. [See Hillenkoetter's statement in this volume.]

60. **Hoover, J. Edgar:** Longtime director of the FBI who made a notation on a memo that confirmed that the US government had recovered crashed disks. [See Hoover's memo in this volume.]

61. **Hynek, Dr. J. Allen:** Perhaps the single most important figure in the study of UFOs. Hynek worked on Project Blue Book, created the Center for UFO Studies (CUFOS), and devised the "Close Encounter" identification system.

62. **Ishikawa, General Kanshi:** The Chief of Air Staff of Japan's Air Self-Defense Force who stated unequivocally that UFOs are real and extraterrestrial in origin. [See Ishikawa's statement in this volume.]

63. **Keel, John A.:** Important UFOlogist and author.

64. **Keyhoe, Major Donald E.:** US Marine and Director of NICAP who was outspoken about his belief that UFOs were real and extraterrestrial in origin.

65. **Klass, Philip J.:** A prominent UFO debunker and the author of *UFO Identified* and *UFOs Explained*.

66. **Lewis, Steve:** A former Air Force intelligence officer who stated that the movie *Close Encounters of the Third Kind* was nonfiction. [See Lewis's statement in this volume.]

67. **Lorenzen, Coral and Jim:** Prominent UFOlogists and authors who investigated the Adamski claims.

68. **MacArthur, General Douglas:** Famous American General who was rumored to have been involved in the establishment of the

Interplanetary Phenomenon Unit. This group was supposedly formed to investigate crashed and retrieved flying saucers.

69. **McDivitt, James:** A US astronaut who admitted a UFO sighting.

70. **McLaughlin, Commander R. B.:** Navy officer who authored an article titled "How Scientists Tracked Flying Saucers" in the March 1950 issue of *True*, which was cleared by the military and which seemed to confirm that UFOs were extraterrestrial in origin and piloted by alien beings.

71. **Mantell, Captain Thomas F., Jr.:** A US National Guard pilot who crashed after pursuing a UFO. His last words were, "It appears to be a metallic object, tremendous in size, directly ahead and slightly above. I am trying to close for a better look." Rumors persist that Mantell was deliberately shot down because of what he saw; that his body was never found; and that his body had strange markings on it.

72. **Marcell, Major Jesse A.:** A 509th Bomb Group intelligence officer who reportedly collected the wreckage of the crashed Roswell saucer.

73. **Marden, Kathleen:** Well-known UFO abduction researcher, author, and lecturer, and Betty Hill's niece to whom Betty told the story of her abduction with husband Barney in New Hampshire on the morning after the incident.

74. **Menger, Howard:** Contactee and author of *From Outer Space to You*.

75. **Menzel, Dr. Donald H.:** A UFO debunker and rumored member of the Majestic 12 group.

76. **Palmer, Ray:** Pioneer UFOlogist and the editor of *Amazing Stories* magazine in the 1940 and 1950s.

77. **Randle, Kevin:** Prominent UFOlogist and author.

78. **Randles, Jenny:** Prominent UFOlogist and author. Randles interviewed Prime Minister Thatcher after the Bentwaters incident. Regarding the Bentwaters UFO sighting and resulting international flap, the Prime Minister reportedly told Randles, "there are some things you just can't tell the people."

79. **Rankin, Richard:** A pilot who sighted ten UFOs in 1947, ten days before Kenneth Arnold had his famous sighting.

80. **Rickenbacker, Captain Eddie:** The World War I flying ace who stated that UFOs were real. [See Captain Rickenbacker's statement in this volume.]

81. **Robertson, Dr. H. P.:** The Chairman of the Robertson Panel.

82. **Ruppelt, Captain Edward J.:** The Project Coordinator for UFO Investigations for the US Air Force in the fifties and the author of *The Report on Unidentified Flying Objects.* Ruppelt coined the term UFO. Captain Ruppelt was the Director of Project Blue Book and was stationed at Wright Patterson AFB during the 1952 Washington, DC, UFO flap.

83. **Russell, Kurt:** Kurt Russell, the star of the film *Stargate* and a licensed private pilot, reported that he was flying into Sky Harbor Airport in Phoenix to drop off his son when he saw the formation of lights at his 12 o'clock. He followed the lights into Phoenix, but forgot about the incident for the next twenty years, only remembering it when he heard a story on the news about the light formation and the small plane that followed the lights into Phoenix.

84. **Sagan, Carl:** Famous astrophysicist and UFO debunker. Sagan was the author of *The Cosmic Connection, Cosmos, The Dragons of Eden,* and other popular books, and based his skepticism about UFOs on the scientific realities of interstellar travel. Sagan did not consider the transdimensional theory of alien travel a viable explanation for how UFOs got to the planet Earth.

85. **Sanderson, Ivan T.:** Biologist, UFO researcher, and Sasquatch (Bigfoot) researcher. Sanderson believes that UFOs come from "other dimensions" and also believes in the existence of other intelligences throughout the universe—both in our dimension and others. Sanderson was one of the first researchers to investigate the incident in the Philadelphia Navy Yard where the destroyer escort USS *Eldridge*, during a degaussing experiment, partially melted, was sailed into a fog down the Intracoastal Waterway, and became the subject of the "Philadelphia Experiment."

86. **Schreiber, Bessie Brazel:** Roswell farmer Mac Brazel's daughter who saw the crashed alien spacecraft debris and who

confirmed the presence of alien hieroglyphics. Bessie was twelve at the time of the crash.

87. **Scully, Frank:** The newspaper columnist who revealed to the world the details of the 1948 Aztec saucer crash and who later wrote *Behind the Flying Saucers.*

88. **Semans, Robert C., Jr.:** The Secretary of the Air Force in 1960 who announced the end of Project Blue Book, primarily because of the findings of the Condon Report.

89. **Simon, Dr. Benjamin:** The Boston psychiatrist who regressed Betty and Barney Hill.

90. **Slayton, Donald:** A US astronaut who admitted to a UFO sighting experienced before he became an astronaut.

91. **Smith, Captain E. J.:** A United Airlines pilot who may have seen the same UFOs that Kenneth Arnold reported in June 1947. Ironically, Smith shared the same name as the captain of the *Titanic.*

92. **Smith, Wilbert B.:** The Canadian scientist who stated that UFOs seemed to be "emissaries from some other civilization."

93. **Spencer, John:** Renowned UFO researcher and author of *The World Atlas of UFOs, The UFO Encyclopedia,* and other books.

94. **Spignesi, Stephen:** Author of *The UFO Book of Lists* and co-author (with Bill Birnes) of *The Big Book of UFO Facts, Figures & Truth.*

95. **Strieber, Whitley:** American novelist and abductee who wrote about his experiences in the best-seller *Communion,* and who continued to discuss UFOs and the extraterrestrial presence on Earth in several follow-up volumes.

96. **Struve, Dr. Otto:** The astronomer who founded Project Ozma and who reportedly picked up intelligent signals from the star Tau Ceti before the project was immediately shut down. Struve was involved in calculating the "Green Bank Formula" which calculated the number of intelligent civilizations capable of communicating with Earth.

97. **Fyfe, Symington:** Governor Fife Symington of Arizona gained notoriety after the March, 1997, Phoenix Lights when, after the media descended upon the Phoenix area when news and video of the formations of lights spread across the news, he held a news conference during which his chief of staff, dressed as an alien, was revealed to the press. Although there were guffaws from the assemblage reporters, those UFO witnesses did not laugh. Years later, Symington apologized for his stunt and revealed that he, too, had been a first-hand witness of the UFO, a huge black translucent triangle that floated over his house on the night of the sightings.

98. **Tesla, Nikola:** Groundbreaking inventor and scientist who received strange radio signals in his laboratory in 1899 which he believed may have been signals from aliens trying to contact us.

99. **Truman, President Harry S:** The American President who issued an official statement confirming that flying saucers were not

constructed by any power on earth. [See Truman's statement in this volume.]

100. **Vallee, Jacques:** UFOlogist and former astronomer, first noted for a defense of the scientific legitimacy of the extraterrestrial hypothesis and later for promoting the interdimensional hypothesis.

101. **Villas-Boas, Antonio:** An early abductee.

102. **Walter, Ed:** The principal witness of the famous 1987 Gulf Breeze, Florida UFO sightings, and the author, with his wife, of *The Gulf Breeze Sightings.*

103. **Walton, Travis:** One of the most famous—and credible—abductees in UFOlogy. His story was the basis for the movie *A Fire in the Sky.*

104. **Welles, Orson:** The producer of the 1938 *War of the Worlds* radio broadcast that started a nationwide alien invasion panic. [See the chapter on Welles's broadcast in this volume.]

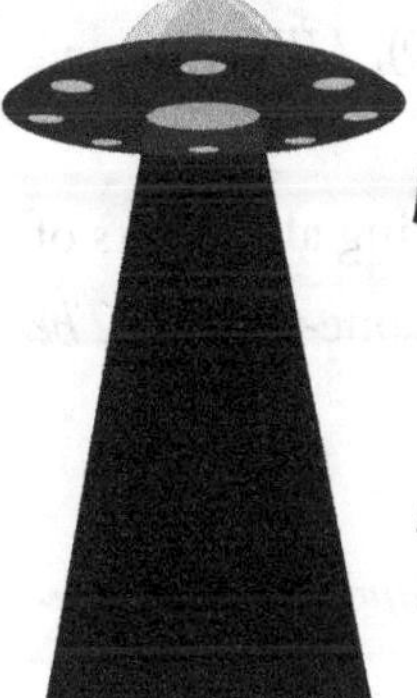

# An Interview with Jerome Clark, author of *The UFO Book: Encyclopedia of the Extraterrestrial*

*What do [reports of UFOs] amount to? We would all be better off if, when the occasion called for it, we pretended to no false authority but instead boldly uttered three one-syllable words seldom heard in the five decades of the UFO controversy: We don't know. Except in those instances where good reason exists to doubt an informant's sincerity,* The UFO Book *operates on the assumption that intellectual agnosticism, rather than its alternatives (occultism on one hand, reductionism on the other), is the wisest course.*

—Jerome Clark, from his Introduction to *The UFO Book*

Jerome Clark, the author of one of the most comprehensive and authoritative books about the UFO phenomenon, is a UFO agnostic.

Clark approaches reports of UFOs and abductions with great skepticism, but also with open-mindedness and an adherence to the scientific method that would make Galileo proud.

Clark has been following the UFO controversy for four decades and is the editor of the *International UFO Reporter*, the official publication of the J. Allen Hynek Center for UFO Studies (CUFOS). He is the author of the award-winning, massive, three-volume UFO Encyclopedia (*UFOs in the*

*1980s, The Emergence of a Phenomenon: UFOs from the Beginning Through 1979,* and *High Strangeness: UFOs from 1960 through 1979*), *UFO Encounters & Beyond, Unexplained!*, and other books.

For those of you serious about balanced reporting covering all aspects of the UFO phenomenon, Jerome Clark's books (*especially* his one-volume *The UFO Book*) cannot be too highly recommended.

**Stephen Spignesi:** *What percentage, if any, of unidentified sightings of aerial phenomena do you believe are of true extraterrestrial origin?*

**Jerome Clark:** The question is meaningless. We don't know that *any* UFO sightings are of "true extraterrestrial origin." Questions about whether there is a core of unexplained UFO sightings are entirely separate from hypotheses about their possible meaning or origin. That's why we call them "UFOs," as opposed, say, to extraterrestrial spacecraft, etheric ships, or whatever.

If you mean how many sightings of ostensible UFOs remain unexplained after investigation, I point to the period between 1951 and 1953, when, under Captain Edward J. Ruppelt, Project Blue Book conducted its most thorough, open-minded investigation—in other words, when the purpose, as it was during most of the project's unfortunate history, was not to debunk reports at any cost. In that period Blue Book found the "unknowns" (as they were called) to comprise 20 to 25 percent of the reports it investigated.

SS: *What are your thoughts about our government's knowledge—or lack of knowledge—of the truth about UFOs and an alleged extraterrestrial presence on Earth?*

JC: The evidence for an official cover-up on some level is not overwhelming, but what evidence exists is undeniably intriguing. Mostly, it consists of a body of testimony from apparently sane, sincere ex-military people recounting sightings they experienced while in service. Some of these alleged sightings are fairly dramatic, involving attempted jet intercepts, photographic evidence, or interference with electrical systems on planes and ships. The informants say that later they were interviewed by unidentified men clad in civilian clothing and warned not to discuss what had happened. The sightings described do not appear in Project Blue Book records. Retired Air Force General Arthur Exon has stated that a secret, well-funded UFO project existed and that the poorly staffed Blue Book was little more than a public-relations exercise.

SS: *Could you describe any UFO sightings that you have personally witnessed? And of the people you have spoken to who have had sightings, what percentage of them do you believe?*

JC: I have never had a UFO sighting, nor have I had any particular desire to have one. With rare exception, people who report UFO sightings are sincere. Hoaxes do occur, of course, but even the Air Force found they comprise 1 percent or fewer of the reports. The problem UFOlogists have to deal with is usually not dishonesty but faulty perception. Overwhelmingly, bogus UFO reports arise out of sincere misjudgments of the phenomenon perceived. People see astronomical bodies, advertising planes, and the like and mistake them for something extraordinary.

SS: *What are your thoughts about the theory that there is evidence of extraterrestrial archaeology on the Moon and Mars, and that these monuments are connected with the terrestrial "ancient astronaut" theory?*

JC: I am thoroughly skeptical of such claims.

SS: *What are your thoughts about the theory that many UFOs are transdimensional rather than intra/intergalactic in origin?*

JC: I find the concept "transdimensional" scientifically meaningless. And given the distances involved—if we assume for the sake of discussion that UFOs may be extraterrestrial in origin—there is virtually no chance they are "intergalactic."

SS: *What percentage—if any—of alleged alien abduction cases do you believe are actual abductions (vs. hoax, psychosis, hallucination, etc.)?*

JC: We don't know that *any* "alien abduction cases . . . are actual abductions."

SS: *If the exact values for all the factors of Frank Drake's "communicative civilizations" equation were actually known, what do you think the results would be? [See the chapter "The Factors of Frank Drake and Carl Sagan's 'Alien Civilization' Equation" for more about this equation.]*

JC: I think, as many astronomers do, that life may be ubiquitous in the cosmos. If this is so, the galaxy is almost certainly densely populated by intelligent, technologically advanced civilizations. If that is so, it is more likely that we would have visitors from elsewhere than that we wouldn't. If, of course, such life is rare, ET visitation becomes correspondingly less probable.

SS: *What do you think of the use of UFOs and aliens in popular culture, i.e.,* Men in Black, *and other movies and books?*

JC: One, such use is inevitable, given the lurid nature of some UFO-related claims. Two, I think it's regrettable. The effect is to marginalize and trivialize the subject even more than it already has been.

SS: *You have researched the UFO phenomenon for decades. Were you initially a skeptic and now a believer? Initially a believer and now a skeptic?*

JC: I have *never* been a "believer." I do, however, consider UFO research eminently worthwhile, and I think that we have much to learn from it. The UFO phenomenon remains a genuine mystery, sadly and unjustifiably neglected—at least so far—by mainstream science.

# An Interview with *Apollo 14* Astronaut Edgar Mitchell

*On the return trip home, gazing through 240,000 miles of space toward the stars and the planet from which I had come, I suddenly experienced the universe as intelligent, loving, harmonious . . . My view of our planet was a glimpse of divinity . . . We went to the Moon as technicians; we returned as humanitarians.*

—*Apollo 14* astronaut, Edgar Mitchell

*Apollo 14* astronaut Edgar Mitchell is a rarity among astronauts: He is the only one of the few who has walked on the Moon to publicly state his conviction that there is an extraterrestrial presence in the universe and that there is a likelihood that we have been visited by alien beings.

Edgar Mitchell (along with Alan Shepherd and Stuart Roosa) was part of the *Apollo 14* team and, on his way back to Earth after walking on the Moon, experienced a moment of spiritual enlightenment which he eloquently describes in the epigraph to this chapter. When he splashed down from his historic mission on February 9, 1971, Edgar Mitchell was a changed man, and this rare epiphany set him on a quest to understand what he describes as the "cosmology of consciousness."

Dr. Mitchell is a scientist, astronaut, naval officer, test pilot, entrepreneur, author, and lecturer. His NASA biography tells us that his "extraordinary career personifies humankind's eternal thrust to widen its horizons as well as its inner soul."

Dr, Mitchell's career is impressive indeed: He has a Bachelor of Science degree in Industrial Management from Carnegie Mellon University, a Master of Science degree from the US Naval Postgraduate School, and a Doctor of Science in Aeronautics and Astronautics from the Massachusetts Institute of Technology (MIT). In addition, Dr. Mitchell has received honorary doctorates in engineering from New Mexico State University, the University of Akron, and Carnegie Mellon. Dr. Mitchell was also awarded the Presidential Medal of Freedom, the USN Distinguished Service Medal, the NASA Distinguished Medal, and three NASA Group Achievement Awards.

After he retired from the Navy in 1972, Dr. Mitchell founded The Institute of Noetic Sciences to study the nature of consciousness. (Author and consciousness researcher John White, who is cited in the Introduction to this book, was Director of Education for the Institute during the seventies.) In 1988, Dr. Mitchell was a co-founder of The Association of Space Explorers, an international organization with a very elite membership: to join, you have to have experienced space travel!

Dr. Mitchell is the author of the 1976 book *Psychic Exploration: A Challenge for Science,* the 1996 book *The Way of the Explorer: An Apollo Astronaut's Journey Through the Material and Mystical Worlds,* and writes regularly for journals and magazines. He is also a highly sought-after lecturer and every year routinely averages one crucial lecture per week.

This interview was conducted on the morning that NASA canceled the Space Shuttle Endeavour flight because of a mysterious alarm in the Shuttle's cabin. Endeavor was the Shuttle flight that was to begin work on the proposed International Space Station. Dr. Mitchell did not comment on the Shuttle flight cancellation, but the metaphorical implications of the event did not escape his interviewer.

**Stephen Spignesi:** *Let's start with this: what percentage of sightings of unidentified aerial phenomena do you believe are of extraterrestrial origin?*

**Dr. Edgar Mitchell:** Well, the question is not really answerable in percentages, but I think that there are very, very few sightings of actual extraterrestrial origin.

SS: *Less than five percent?*

EM: I don't want to put a number on it because I really don't know, but it's somewhere in that vicinity, I'm sure.

SS: *What are your thoughts about our government's knowledge of the truth about UFOs and the extraterrestrial presence on Earth?*

EM: Well, I have to qualify the word "government." At one time the government, and by that I mean high-level government, was quite in on it. The documentation coming out now is very, very compelling and, at that earlier time, there was probably justification for such secrecy. There is no longer justification for a cover-up regarding such matters. I also think most people in government are still quite naive about all this, but there is a faction in our government—a "quasi-government"—that is well aware of the truth.

SS: *To follow up on that, there has been talk that one of the reasons why the truth has not been not revealed for so many decades is because the Orson Welles's "War of the Worlds" panic demonstrated to global governments that "the masses" were not capable of handling the truth about a possible alien presence elsewhere in the universe and possibly on Earth. Your thoughts about that?*

EM: That's a good rationalization, and perhaps at that time global panic was a bona fide fear, but there's been plenty of time in the last fifty or sixty years to adequately prepare the people for the truth, and I think overall, the population is more than well-prepared for it.

SS: *Could you describe any UFO sightings that you have personally witnessed in your life?*

EM: None. None at all. I keep making this point over and over again. I have no firsthand experience with either sightings or extraterrestrial contact. What I have done is relied most heavily on government people of a previous period—people I call the old-timers—who wanted to tell their story before they died and have said, "Yes, all of this is true." And their assertions are all bolstered by documentation that's coming out that seems to be impeccably researched and which augments and bolsters the claims of the other documents that have been leaked out over the years and debunked. But it now looks like the evidence is very convincing.

SS: *As one of the few humans who have ever walked on the Moon, your answer to this next question has great significance. What are your thoughts about the theory that there is evidence of extraterrestrial archaeology on the Moon and Mars?*

EM: All the evidence I've seen is not persuasive at all. I see nothing that substantiates that theory, particularly on the Moon. That isn't to say it couldn't be true, but I don't think that the evidence we've seen is at all compelling.

SS: *What are your thoughts about the recent theory that many UFOs are transdimensional, hyperdimensional, or multidimensional, rather than intra- or intergalactic?*

EM: I don't think that's true at all. I think modern work in physics is tending to confirm that the universe is a three-dimensional space universe and a four-dimensional space/time universe—just like it appears to be. And even though these multidimensional theories are the current rage in mathematics and in some areas of physics, particularly those attached to string theory, I think that the test of time will show that the universe is really just a three-dimensional universe. Also, I think that we will ultimately discover that most of the numinous effects that we perceive in the universe are associated with quantum nonlocality.

SS: *What percentage of alien abduction cases do you believe are actual abductions vs. hoax, psychosis, hallucination, or something else?*

EM: It's hard to put a number on it. Again, we have to look at the evidence. One can say that those cases that have substantial, verifiable, physical evidence associated with them could very well be true physical abductions. And there are a few of those. But I've often expressed the view that this is a considerably more complex phenomenon. We're dealing with the realm of people's perception; their interpretation of perceptions, and so, the truth ranges across the whole panoply of explanations that you just described: Everywhere from the collective unconscious to bad dreams to hoax and beyond. Even well-meaning people can misinterpret their experiences. So there may be some small percentage that are true physical abductions, but it is nowhere near the roughly two percent of the population that claims them.

SS: *In your opinion, what percentage of NASA astronauts believe that UFOs are of extraterrestrial origin?*

EM: I think most people today, including many NASA astronauts, are now beginning to realize that there probably is an extraterrestrial presence throughout the universe. The question of, have we been visited, though, is quite a different question. The one thing that the astronauts have talked about and agreed upon is that, through the *Apollo* and the

*Skylab* programs, none of us have ever had any encounter in space with unexplainable objects. Generally, the thought processes and belief systems of astronauts parallel those of the general population, so I would say probably more astronauts believe than fifteen years ago, and perhaps not as many as those that will think that way ten years from now. But I don't have hard numbers. I can tell you, though, that the only thing we've really discussed in depth among ourselves—to make sure we're all in agreement—is that, through the Skylab program, none of us have had what I prefer to call "unexplainable encounters" in space.

SS: *If the true values of all the factors of the Frank Drake "communicative civilizations" equation were known, what do you think the results would be?*

EM: My answer to that is that it is likely that life has begun on virtually every planet where the environmental conditions permit; and that intelligent life has certainly evolved wherever the environmental conditions permit. Which would mean that the answer to your question is that there may be billions and billions of "communicative civilizations" throughout the universe.

SS: *What do you think of the use of UFOs and aliens in popular culture, i.e.,* Men in Black, *and other UFO-themed movies and books?*

EM: I think it prepares people to accept the fact that this is a possibility. I do regret some of the wild fringe stuff that comes out of Hollywood, but as long as we recognize that their job is to tell stories—things to stir the emotion—that's fine. But serious people are going to see beneath all that and look for the hard evidence. Unfortunately, there are not enough serious people who are looking at the issues and asking, what's verifiable, what's the evidence, and what does science have to say about it. Regrettably, mainstream science has remained terribly silent on all of this for years, when they shouldn't have.

SS: *Many UFOlogists have claimed that Steven Spielberg's movie* Close Encounters of the Third Kind *should have been released as a documentary. Your thoughts about that?*

EM: *Close Encounters* was based upon J. Allen Hynek's work. Hynek was a very serious and thorough investigator whom I knew personally, and the evidence would suggest that there has, indeed, been that type of contact.

SS: *What are you working on?*

EM: I am continuing my three decades of work on consciousness studies. Primarily, I am interested in such questions as, how did consciousness evolve, and what is the nature of the universe that would permit that to happen? I now think we're getting close to a good answer to those questions. Research in holography and quantum holography seems to indicate that the universe—bottom line—is an intelligent, creative, self-organizing, trial-and-error learning system. I will be coming out with a new book about all this when I finally get it all pulled together.

SS: *Dr. Mitchell, it has been an honor having you in this book, and I sincerely appreciate your participation.*

EM: Thank you.

## EPILOGUE

From the *Associated Press*, October 15, 1997

**PHOENIX**–*A former astronaut who walked on the moon says he thinks aliens have crash-landed on Earth. Edgar Mitchell, the sixth man to walk on the moon, said he believes some military and other planes use* *technology derived from alien spacecraft that have been captured and disassembled. The purported secret project has been going on for decades under a parallel government administration, separate from the president and the highest-ranking members of the Pentagon, Mitchell said. On Saturday (10-11-97), Mitchell called for congressional hearings on whether the United States has captured alien craft and studied them to produce new technologies.*

## POST-MORTEM

Edgar Mitchell died under hospice care in West Palm Beach, Florida, on February 4, 2016, at the age of eighty-five.

# The Condon Committee's "UFO Opinion" Questionnaire

On October 7, 1966, the United States Air Force commissioned the University of Colorado to conduct a serious inquiry into UFOs.

The project was helmed by the renowned scientist Dr. Edward U. Condon and cost the American taxpayers close to $500,000. The project became known as the Condon Committee, and its findings were released to the public on January 8, 1969.

As part of its research, the Condon Committee circulated a UFO Opinion Questionnaire consisting of twenty-nine statements. For each statement, the participant had to indicate the degree to which he or she agreed or disagreed with the statement.

Participants were given four options for evaluating *their* view of the statement:

- **Definitely false:** You are completely convinced that the statement is false and you would act without hesitation in your belief. You would not be able to be swayed from your belief by another person's views on the statement.
- **Probably false:** You are not really sure the statement is false, but if forced to choose, you would lean towards the statement being untrue. Your opinion might be able to be changed by someone else's views.

- **Probably true:** You are not really sure the statement is true, but if forced to choose, you would lean towards the statement being accurate. Your opinion might be able to be changed by someone else's views.
- **Definitely true:** You are completely convinced that the statement is true and you would act without hesitation in your belief. You would not be able to be swayed from your belief by another person's views on the statement.

The following list consists of the twenty-nine Condon Committee UFO statements. You might want to take this questionnaire yourself and then evaluate your answers. Upon analysis you will probably find that you fall into one of the following four categories:

- **Unflinching disbeliever:** The idea of UFOs and extraterrestrial beings is a non-issue with you: They simply do not exist, and you feel strongly that anyone who believes they do is delusional.
- **Skeptical agnostic:** You're really not sure if UFOs are real or if we have been visited by aliens from other planets, but deep down you tend to be a doubting Thomas, wondering why, if alien contact is so common, there isn't evidence so undeniably concrete that there could no longer be any doubt about the existence of UFOs and ETs.
- **Open-minded agnostic:** You're really not sure if UFOs are real or if we have been visited by aliens from other planets, but deep down you ascribe to what Thomas Carlyle once wrote in a letter: "I don't pretend to understand the Universe—it's a great deal bigger than I am . . . People ought to be modester."
- **Passionate believer:** Of course, UFOs are real. We have been visited by beings and spacecraft from many other planets and dimensions for millennia; the government knows it,

and it's just a matter of time before man's consciousness evolves enough to recognize our neighbors and join with them as part of a universal consortium of peaceful beings.

## Condon Committee statements for evaluation

1. Some flying saucers have tried to communicate with us.
2. All UFO reports can be explained either as well understood happenings or as hoaxes.
3. The Air Force is generally doing an adequate job of investigation of UFO reports and UFOs.
4. No actual, physical evidence has ever been obtained from a UFO.
5. A government agency maintains a Top Secret file of UFO reports that are deliberately withheld from the public.
6. No airline pilots have seen UFOs.
7. Most people would not report seeing a UFO for fear of losing a job.
8. No authentic photographs have ever been taken of UFOs.
9. Persons who believe they have communicated with visitors from outer space are mentally ill.
10. The Air Force has been told to explain all UFO sightings reported to them as natural or man-made happenings or events.
11. Earth has been visited at least once in its history by beings from another world.

12. The government should spend more money than it does now to study what UFOs are and from where they originate.
13. Intelligent forms of life cannot exist elsewhere in the universe.
14. Flying saucers can be explained scientifically without any important new discoveries.
15. Some UFOs have landed and left marks in the ground.
16. Most UFOs are due to secret defense projects, either ours or another country's.
17. UFOs are reported throughout the world.
18. The government has done a good job of examining UFO reports.
19. There have never been any UFO sightings in Soviet Russia.
20. People want to believe that life exists elsewhere than on Earth.
21. There have been good radar reports of UFOs.
22. There is no government secrecy about UFOs.
23. People have seen spaceships that did not come from this planet.
24. Some UFO reports have come from astronomers.
25. Even the most unusual UFO reports could be explained by the laws of science if we knew enough science.
26. People who do *not* believe in flying saucers must be stupid.
27. UFO reports have not been taken seriously by any government agency.
28. Government secrecy about UFOs is an idea made up by the newspapers.
29. Science has established that there are such things as "Unidentified Flying Objects."

Alas, although the individual scientists contributing to the Condon University of Colorado Report believe that there existed tangible evidence of UFOs, because Dean Condon had been compromised, his introduction to the report, disputing some of the findings in the report, dismissed the question of UFOs so that the Air Force could shut down Blue Book.

# PART IV
# ALIEN ANCESTRY

# We Are the ETs

*Our quest on Mars has been to "follow the water" in our search for life in the universe, and now we have convincing science that validates what we've long suspected.*

—John Grunsfeld, Astronaut and Associate Administrator of NASA's Science Mission Directorate in Washington

In David Bowie's seminal 1971 album *Hunky Dory*, his timeless song "Life on Mars" asked the question, "Is there life on Mars?"

What would it mean if there was conclusive proof that life at one point existed on Mars? Is Mars now what the Earth will look like in a billion years? Did life on Earth come from Mars? Were we seeded?

## Have We Met ET?

That is the question that has puzzled UFO enthusiasts ever since the stories of ETs climbing out of the crashed delta-shaped craft at Roswell. Abductees have claimed to have seen both small grays and large grays, Nordics, and Reptoids, but, according to planetary scientists, the real mystery surrounding the identity of ET is right here on Earth.

Shortly after the formation of our solar system, planetary scientists theorize, our planet was not only pummeled by meteors, we were also pummeled by chunks of other planets, specifically large hunks from Mars, blown off the red planet by its bombardment of meteors. The impact on our planet was so great that a huge chunk of Earth was blown off and became trapped in orbit. And that's one plausible theory accounting for the existence of our moon.

But, as scientists have also found, it gets more complicated. Because our universe is composed of the same material throughout, heavy in elements like hydrogen, carbon, silicon, and oxygen, there is a theory that if there is life out there, it is probably composed of much the same elements that compose life on earth. Maybe life out in our galaxy doesn't look like life on earth, but, chemically, the theory goes, what comprises us, comprises life out there.

But it even gets more complicated, because if Earth is a big chemical crockpot, something had to start the life process going. That process may well have come from meteors that hit our planet and dropped into our oceans to create a form of primordial soup. Imagine a recently formed Earth holding deep oceans bubbling with gasses spewing up from deep volcanoes. Into this soup, microbe-carrying meteors begin to deposit the fundamentals of self-replicating microbes, the tiniest forms of life that begin to adapt themselves to this new environment and begin to evolve.

Or, imagine further, that what we hypothesize what happened on earth also happened on Mars. At an earlier stage of its existence, Mars, we believe, had an atmosphere. Imagine further that the canyons and gorges we have spotted on Mars also held liquid water. If what we theorize is true, at a very early stage, Mars might well have hosted life. If at that stage of a life-hosting Mars, chunks of Mars broke off from the planet and plummeted into Earth's oceans, that, too, might have stimulated the evolution of life on Earth.

Supposing this is true, and microbial life on Mars did reach Earth, it's conceivable that we're actually not Earthlings.

We are the ETs.

# The Factors of Frank Drake and Carl Sagan's "Alien Civilization" Equation

*"While the evidence for aerial craft visiting Earth goes back as far as recorded history—and further if cave painting etc. are taken into account—the modern wave of UFO sightings suggests that it is human beings' acquisition of advanced technology which has spawned a probably unprecedented interest in us and our welfare by space races from many different parts of the galaxy."*

—Alan Watts

## UFO Quest: In Search of the Mystery Machines

In 1961, the two theoretical astronomers Frank Drake and Carl Sagan devised a mathematical equation that, if solved, would indicate the number of "communicative civilizations" within the Milky Way Galaxy.

That equation:

$$N = R^* x\ f/p\ x\ n/e\ x\ f/l\ x\ f/l\ x\ f/c\ x\ L$$

The only problem, however, with this equation is that none of the values of the factors of the equation are known with certainty. (Even R*, the actual number of stars in the Milky Way, is not precisely known, although 150 billion is now the number science uses on a regular basis.) Thus, the momentous value "N"—the actual number of civilizations in our Galaxy—must, by necessity, remain elusive.

And yet, educated guesses can be made as to what values to plug in to the equation's factors to come up with an estimated answer.

Here is what each of Sagan's and Drake's factors stand for:

- **N**—*the solution to the equation*—is the number of civilizations in the Milky Way Galaxy advanced enough to communicate with other similarly developed civilizations (like us).
- **R*** is the number of stars in the Milky Way Galaxy.
- **f/p** is the fraction of the stars in the Milky Way Galaxy that possess planetary systems.
- **n/e** is the average number of Earthlike planets within the planetary systems in the Milky Way Galaxy.
- **f/l** is the fraction of Earth-like planets where life develops.
- **f/i** is the fraction of these living organisms that evolve into intelligent life-forms.
- **f/c** is the fraction of these intelligent species who are able and willing to attempt communication with other similarly evolved life-forms on other planets capable of receiving messages.
- **L** is the average lifetime of an intelligent civilization expressed as a fraction of the lifetime of the Milky Way Galaxy.

For those of you who would like to try plugging some numbers into this equation and seeing how many intelligent civilizations *you* come up with, I can give you a couple of facts with which to work. (And by the way, when astronomer George Abell plugged *his* best guesstimates for the different factors into the equation, he came up with between 100 million and 10 billion civilizations capable of communicating with us—and that's just in the Milky Way Galaxy. That sure is a lot of intragalactic chatting, isn't it?)

For the factor R* (the number of stars in the Milky Way), you can use the figure most often cited, 150 billion (150,000,000,000).

And for the factor L, the average lifetime of an intelligent civilization expressed as a fraction of the lifetime of the Milky Way Galaxy, we know that the Milky Way is now approximately 12 billion years old (12,000,000,000) and our own Earth civilization has now existed for 40,000 years. It's up to you to come up with an estimate as to how long our intelligent civilization will continue to survive, as well as numbers for all the other factors.

Okay, now you're ready to do your math! (You did pay attention in school during algebra class, now didn't you?)

Happy postulating!

## The Ancient Alien Theory

The ancient alien theory posits that hundreds of thousands of years ago, extraterrestrials came to planet Earth where they manipulated the hominid DNA to perfect a species that would serve their purposes, specifically mining for the precious metals they required.

Alternatively, the theory suggests, ancient aliens also sought DNA samples from early hominids to sustain their own species. The theory goes on to suggest that over the course of the hundreds of thousands of years, these ancient extraterrestrials involved themselves in the development of human cultures, creating what we now call "cargo cults" requiring human civilizations to obey laws ostensibly handed down by deities. For evidence of this ancient extraterrestrial involvement in human development, theorists have pointed to artifacts, such as the Egyptian pyramids, which, theorists argue, could not have been constructed given the construction technology at the time.

Other artifacts also have captured the attention of ancient alien theorists, including sites such as Machu Picchu in South America, where the stones had to be brought up to the tops of a mountain and then hand hewn so that they perfectly fit. Other examples include the Nazca Lines, straight lines laid out in such a way that they could only be seen from the sky.

One of the proponents of the ancient alien theory was Erich von Daniken, whose 1968 book, *Chariots of the Gods*, inspired a 1970 television movie and spurred research into the artifacts von Daniken described as having been constructed, if not directly by aliens, then certainly with alien help.

There have been decades of pushback against von Daniken's theory, especially from a number of skeptics, who argue that, first, ancient civilizations, especially the Egyptians and the Aztecs and Incas, were certainly capable of harnessing human power to build their respective structures. They also argue that because there are plausible conventional explanations for ancient artifacts that seem too large or complex for the civilizations at the time, it is too much of a jump to argue that extraterrestrials were responsible for them.

However, ancient alien theorists also argue that the literature and culture of what we believe to be ancient civilizations also point to a pre-civilization even before what we consider to be Stone Age cultures. They specifically cite a structure like the temple at Gobekli Tepe in Turkey's Anatolia region, even though it has been commonly considered to be an early spiritual sanctuary for Neolithic society. However, and this is one of the most puzzling aspects of this architectural artifact, even the early inhabitants of ancient Sumer believed that Gobekli Tepe was constructed by what they call "the old ones."

Who were the old ones?

The question is also posed by scientists who argue that the Sphinx shows evidence of water erosion rather than wind erosion, which is significant because the Sphinx sits in a desert climate. Where would the water have come from to erode the Sphinx? If there is evidence of early water erosion on the Sphinx, it means that a certain point in history, what was desert was more tropical.

Part of the ancient alien theory was also posited by Immanuel Velikovsky, whose 1950 book, *Worlds in Collision*, argued for the "electronic universe." The book also suggested that the events set forth in the first five books of the Bible, such as the plagues in Egypt and the parting of the Red Sea, were not the result of miracles, but natural occurrences stemming from the proximity of another planet that came so close to the Earth, its magnetic pull affected life on Earth.

Ancient alien theorists also argue that one of the reasons for the UFO cover-up, particularly in the United States and Western Europe, is not just the domination of the Church but the fear that if the truth were told about the involvement of ancient visitations by extraterrestrials and their manipulation of the human genome, world religion would collapse along with world government.

However, the question remains, did humans naturally evolve? Or was our species created by intelligent design and is still being manipulated by extraterrestrials today for some larger purpose?

PART V

# TAKEN: ABDUCTIONS BY ALIENS

# Baffling UFO Abduction Cases

Forcible abductions by aliens have been reported for decades, and a majority of the stories told by the victims share similar elements, including the descriptions of the beings, the descriptions of the ships, and the descriptions of the tests and procedures performed on the abductees.

Here is a look at six of the most notable abduction cases of the past three decades.

### Antonio Villas-Boas (October 16, 1957)

This fascinating Brazilian abduction case has the added element of interspecies sex, specifically Villas-Boas's claim that he had sexual intercourse twice with a shapely naked alien female who had slanted eyes, white hair and who growled like a dog during the sex act.

Villas-Boas claimed that he was taken aboard a spacecraft where beings in spacesuits undressed him, took a blood sample from him, and then covered his body with a thick, odorless liquid.

He was then left alone and shortly thereafter, the naked alien woman came in, manually aroused him, had sex with him twice, and then stored some of his sperm in a container before exiting.

Villas-Boas also claimed that before she left the "woman" pointed to her belly and then towards the sky and he interpreted this to mean that he had impregnated her and that their child would be born somewhere in outer space.

Villas-Boas ultimately became a lawyer and had four children. He recently died without recanting his story. Many UFOlogists consider Villas-Boas's story credible and point to his lifelong refusal to profit from his story and his upstanding reputation as supportive factors.

### Betty and Barney Hill (September 19–20, 1961)

The Hill Abduction case is the most famous UFO abduction case to date.

In 1966, journalist and UFO investigator John G. Fuller (now deceased) published a book about the Hill case called *The Interrupted Journey*. At the conclusion of his book, Fuller wrote, "An abduction by humanoid intelligent beings from another planet in a space craft [sic] has always belonged to science fiction. To concoct a science fiction story of this magnitude would require an inconceivable skill and collaborative capacity. It is as hard for the Hills to accept that the abduction took place as it is for any intelligent person. In fact the attitude of the Hills is: We did not expect or look for the sighting to take place. Barney resisted and personally tried to deny its existence. We did not know what happened in the missing two hours and thirty-five miles of distance . . ."

Here is a brief summary of the Hill case facts:

On their way home from a vacation trip in Canada, Betty and Barney Hill, an interracial couple who were unanimously thought of by their neighbors and friends as sober-minded, hard-working, solid citizens, saw a bright light in the sky that paced their car as they drove through New Hampshire. According to details uncovered through hypnotic regression of the Hills two years later, this mysterious light was actually an alien spaceship. Hypnosis revealed that the Hills were taken aboard this craft and underwent two hours of medical examinations, including the taking of a semen sample from Barney, and skin scrapings from both.

After being "instructed" not to remember any of what had happened to them, the Hills were released from the craft, and proceeded home, where they realized that a four or five-hour journey had taken seven hours.

After extensive interviews with the Hills and review of the tapes of their hypnotic regression sessions with Boston's Dr. Benjamin Simon (an avowed UFO skeptic), John Fuller came to the following eight conclusions [See *The Interrupted Journey* for Fuller's detailed explication of these eight findings]:

1. A sighting of some kind took place.
2. The object sighted appears to have been a craft.
3. The sighting caused a severe emotional reaction.
4. The anxiety and apprehension engendered by Barney Hill's racial sensitivity served to intensify the emotional response to the sighting.
5. The Hills had no ulterior motive to create such a story. They had confined their experience to a small group of people for four years.
6. The case was investigated by several technical and scientific persons who support the possibility of the reality of the experience.
7. There is a measurable amount of direct physical circumstantial evidence to support the validity of the experience.
8. Under hypnosis by a qualified psychiatrist, both the Hills told almost identical stories of what had taken place during their period of amnesia.

The Hill case is also buttressed by the fact that Betty Hill drew from memory a "star map" that she remembered under hypnosis as having seen aboard the spacecraft. In 1968, an amateur astronomer named Marjorie Fish deciphered Betty's map and concluded that the origin of the aliens had been the pair of stars Zeta 1 Reticuli and Zeta 2 Reticuli, which are less than fifty-four light years from Earth.

The Hill case is still a mystery and, as UFOlogist Jerome Clark noted in The UFO Book, "The resolution of the Hill case awaits the resolution of the UFO question itself. If UFOs do not exist, then Barney and Betty did not meet with aliens. If UFOs do exist, they probably did."

### Debbie Tomey ("Kathie Davis") (1966–present)

Debbie Tomey was the subject of Budd Hopkins' important 1987 book, *Intruders: The Incredible Visitations at Copley Wood*, and according to Hopkins, Tomey has repeatedly been abducted by aliens since the age of six.

Hopkins came to the conclusion that Tomey was part of an extraterrestrial genetic study and that she may even have been artificially inseminated by aliens at one point. This human/alien hybrid was then removed from Tomey's body and later, she was allowed to interact with nine babies that may or may not have been human.

Tomey, who recently went public with her real name, fully supports the work of serious UFO investigators and credits Hopkins' willingness to believe her as the reason she did not commit suicide after she realized what had been happening to her for so many years.

### Betty Andreasson (January 25, 1967)

Andreasson claimed that on January 25, 1967, aliens entered her home in Ashburnham, Massachusetts through the walls. She was then abducted by a being named Quazgaa, medically examined, and claims that she heard the "Voice of God."

During hypnotic regression, Betty remembered having a needle probe inserted in her navel (almost exactly like the procedure Betty Hill remembered having performed on her) and also recalled having something removed from her nostril. This incident suggested to UFOlogists that the aliens had been monitoring Betty with some kind of high-tech probe, and there is now an entire field of study that looks at alien implantation of probes and their surgical removal.

## Travis Walton (November 5, 1975)

The Travis Walton case is one of the most well-known modern abduction cases and, in the almost twenty-five years since the incident, none of the principals involved have stepped forward (for what would be significant financial gain) to reveal the abduction as a hoax.

On Wednesday evening, November 5, 1975, Travis Walton and six other members of a wood-cutting crew were driving home after a long day clearing trees in the Apache-Sitgreaves National Forest in Arizona. The seven men had been working on a contract for the US Forest Service and because they were behind schedule, had worked until dark.

As they drove through the forest in foreman Mike Rogers' pickup truck, they all suddenly saw a glow emanating through the trees ahead of them and to their right. As they continued slowly down the rough road, the source of the light came into full view: it was a 20-foot wide by 8-foot high airborne disk that had geometric panels surrounding its perimeter and an outer ring encircling its middle.

Travis Walton felt a sudden urge to see the UFO at close range. He later admitted that he immediately knew that this was the "chance of a lifetime" and did not want to miss it.

Travis ran into the clearing, eventually standing directly below the craft, while the others remained in the truck, shouting at him to come back.

As the men in the truck watched in horror, a blue beam enveloped Travis and lifted him up into the air. At first, it seemed as though the beam could not carry Travis's weight because after rising slightly, he was suddenly dropped back to the ground. Upon seeing this, Mike Rogers and the others fled in the truck. They ultimately returned, but there was no sign of Travis, and they all reported seeing a white streak fly over them as if the UFO had suddenly taken off—with Travis inside it. After a perfunctory twenty-minute search, the men returned to town and reported the incident to the police.

The police searched the area over the next few days but to no avail. Then, five days after the incident, Travis's brother-in-law Grant Neff received a phone call from Travis, who told Neff that he was hurt and needed help. Neff found Travis in a phone booth at a service station on the outskirts of Heber, Arizona. He was dehydrated, had five day's growth of beard, and seemed to be in shock, but otherwise was unharmed.

Travis Walton later told a story familiar to those versed in abduction accounts. He had been taken aboard a spacecraft manned by beings dressed in one-piece jumpsuits who did not speak:

> *They were short, shorter than five feet, and they had very large, bald heads, no hair. Their heads were domed, very large. They looked like fetuses. They had no eyebrows, no eyelashes. They had very large eyes—enormous eyes—almost all brown, without much white in them. The creepiest thing about them were those eyes.*

Walton remembered walking through the ship and encountering other more "humanoid-like" beings and then being placed on an examination table. A clear mask was placed over his nose and the next thing he knew he was in Heber. He remembered seeing the craft fly away after "dropping him off" and that was when he called Neff.

Polygraph exams were conducted on all the participants in the encounter, and all of the men were determined to be telling the truth. UFO debunker Philip J. Klass made a career out of telling the world that Walton's and the others' stories were all part of an elaborate hoax, and yet not one of the men involved ever sold their stories, aside from a tabloid article shortly after the incident.

In 1978, Walton published *The Walton Experience* about his encounter, and 1993, the movie *Fire in the Sky*, which was based on Walton's book, was released to modest success. In 1996, Walton published an expanded, revised edition of his book called *Fire in the Sky: The Walton Experience*, which addressed all of the debunking accusations and charges of hoax that had piled up over the twenty or so years since he had had his experience.

Walton has since appeared at many UFO conventions, talking about his abduction, and most who hear him speak agree that he consistently comes across as credible and sincere.

### Whitley Strieber (December 1985)

I was working on my *Complete Stephen King Encyclopedia* in 1987 when Whitley Strieber published *Communion*, which told of his ongoing contact with, and abduction by aliens. Before this nonfiction account, Whitley had published mainly in the science fiction and horror genres, and he and I shared many mutual contacts and colleagues in the field. I spoke to several people who knew Whitley very well and—*Communion's* instant best-seller status notwithstanding—every one of them believed that Whitley had had an extreme paranormal experience and that if Whitley said he had been abducted by aliens, then he had been abducted by aliens.

Admittedly, one or two horror writers who knew Whitley suspected that he had made the whole thing up to give his career a boost, but the vast majority of industry pros held Whitley in very high regard and refused to believe that *Communion* was a publicity stunt.

*Communion* chronicled more than just a single abduction case. In his book, Whitley told of a lifetime of contact with extraterrestrials, beginning in childhood and continuing through his adult life.

"The abduction experience is comparable to being run over by a train," he said. "Your whole life changes. But I guess I'd rather have the abduction experience because you can survive that."

# 9 Techniques to Defend Yourself Against Alien Abduction

*[Alien abduction] behavior clearly violates our own personhood. My acceptance of them is entirely conditional and must be earned. How? By demonstrating that they respect human beings as having inalienable rights. It's nothing more than the Golden Rule. If they won't respect our personhood and our property, then they're predators and deserve whatever means we can take to protect ourselves.*

—John White, quoted in *How to Defend Yourself against Alien Abduction*

Ann Druffel, author of the 1998 book *How to Defend Yourself against Alien Abduction* (Three Rivers Press), began investigating UFOs in 1957 with the National Investigations Committee on Aerial Phenomena (NICAP) and has spoken on and written widely about the UFO phenomenon in magazines, journals, and books.

In *How to Defend Yourself against Alien Abduction*, Druffel details the specific techniques that can be used to protect oneself from alien kidnapping and explains how to terminate an abduction safely. The following list is a summary of the techniques Ann Druffel recommends.

[Special thanks to Ann Druffel and John White for their assistance with this chapter.]

1. **Mental struggle:** This is often the first defensive technique used against abduction, and it is usually instinctive and reactive in nature. Druffel states that the Mental Struggle defense involves "sustained willpower" on the part of the targeted witness as he or she attempts to move a finger or a toe and break the paralysis that is often the first stage of assault by an extraterrestrial abductor. The witness must feel a sense of outrage at being attacked in such a manner and maintain a strong conviction that the Mental Struggle technique will work—even though they are terrified. (Many witnesses who foiled abduction attempts reported that they sensed total surprise on the part of the aliens that the human they were trying to abduct would actually resist!) The targeted abductee must also be patient in applying Mental Struggle: Druffel states that this technique requires "strong, internal, silent struggle" to be effective.

2. **Physical struggle:** This technique exploits the human being's natural instinct to resist an attacker physically, and requires feeling a strong sense of violation. Druffel tells us that this is most effective if applied *before* physical paralysis sets in and that weapons can often be helpful in boosting the victim's confidence that he or she can fend off the abduction attempt. Anger can also be a valuable substitute for self-confidence.

3. **Righteous anger:** This technique is often most effective after repeated abduction attempts and before paralysis. Druffel recommends using strong commands, either verbal or mental, against the attacker, such as "Go away!" or "Leave me alone!" In order to be most effective, righteous anger should

be focused on reclaiming the victim's rights to be left alone instead of hatred toward the alien attacker.

4. **Protective rage:** This technique is used to assert protection for family members, such as small children and others who are virtually defenseless against attack. Techniques that can be used are mainly verbal and include loudly spoken retorts and also profanity against the attackers. It is best used before paralysis, and Druffel emphasizes that it should be used in conjunction with a "thought message" to the attackers that you do not wish them harm, but that you reject their attack and will defend your home and family from them. Interestingly, this technique can also be used after the aliens depart.

5. **Support from family members:** This technique involves willing support and assistance from people emotionally bound to a possible abductee—"family" in this context does not just mean blood relatives—and efforts should focus on protection from, and defense against alien attacks. Druffel emphasizes that this will not work with family members or friends who feel that their assistance should be directed toward "curing" the victim of their apparently delusional thinking. People willing to help must accept that the victim is being attacked and do everything he or she can to put an end to the assaults.

6. **Intuition:** This technique essentially requires potential abductees to pay attention to their intuitive "sense of danger" and prepare themselves accordingly. Often, aliens will not approach if they sense that their victim is mentally and/or physically ready to defend him- or herself against attack.

7. **Metaphysical methods:** This technique involves visualizing a bright white light flowing down through the top of your head, all the way through you, and extending a few inches out away from your body. The white light acts as a psychic shield against attack and is a very effective "alien deterrent."

8. **Appeal to spiritual personages:** This technique involves praying to spiritual beings and asking for their protection. Guardian angels, Jesus, Mary, saints, departed loved ones, and other ethereal beings often prove to be stalwart defenders against evil attacks.

9. **Repellents:** This technique uses physical totems as defense tools against alien attack and possession by demons and succubi. Commonly used objects include crucifixes, holy medals, crosses, herbs, oils, salt, and magnets.

# 24 Medical Procedures Performed On UFO Abductees

*Temple University historian David Jacobs has . . . defined the basic reported pattern of an abduction experience. Jacobs identifies primary phenomena such as manual or instrument examination, staring, and urological-gynecological procedures; secondary events, including machine examination, visualization, and child presentation; and ancillary events, among them miscellaneous additional physical, mental, and sexual activities and procedures.*

—John Mack, *Abduction: Human Encounters with Aliens*

Some of the most bizarre aspects of alien abduction cases are the abductees' reports of "medical" procedures performed on them by their captors.

Often retrieved during hypnotic regression, the abductees' tales are specific, graphic, and often extremely distressing to recall.

Here is a look at the two dozen most commonly cited procedures performed on humans by their alien abductors.

1. **Anal Probes:** Some abductees have reported having an instrument inserted into their rectum and then feeling something cold and hard crawling all throughout their lower abdomen. An abductee named Dave, talking to John Mack in

*Abduction* said that a male alien doctor "[put] this thing, it's a couple of feet long, sort of looks like a thing that they root out sewer pipes with. It sort of has like a large end on it, sort of wiry or something like a wire structure-type thing on the end of it." Dave told Mack that they put the thing up "a lot farther than you could believe it would go." Dave reported that he felt "discomfort and humiliation but little pain" and described feeling "like a zoo animal."

2. **Semen/Egg Extraction:** Barney Hill reported having a suction-like device placed over his groin and feeling a painless extraction. A circle of warts later grew on his skin where the cup had been placed. Some female abductees report a similar device being placed over their vaginal opening and feeling the same kind of extraction. One male abductee cited in John Mack's *Abduction* reported having a tube placed over his penis and needles attached to wires placed on his testicles and his neck. The victim believed that electrical stimulation of these needles caused him to attain an erection and ejaculate into the tube. Other male abductees have reported actual surgical extraction of semen through an incision in the scrotum.

3. **Physical Examination:** Many abductees report being physically touched and palpated by alien doctors, all of whom had ice-cold hands. The aliens counted vertebrae, felt the bones of the limbs, and ran their hands over the abductees' body.

4. **Tissue Samples:** Betty Hill and most other abductees who were examined by alien doctors said that hair and nail samples were taken from them during their examinations by the alien abductors.

5. **Needle Aspiration:** Betty Hill remembered an enormous needle being painfully inserted into her navel during her examination. When she cried out in pain, an alien waved a hand in front of her face, and the pain vanished immediately. She was told that this aspiration was a "pregnancy test." Betty did not believe this and intuitively sensed that they were extracting her eggs.

6. **Skin Scrapings:** Some abductees report having a blade-like instrument scraped over their skin and then watching as the aliens placed the scrapings on glass slides.

7. **Skin Probes:** Some abductees report having an instrument (usually round- or square-shaped) placed against their skin and then moved slowly across the surface. This device seems to be some kind of computerized probe and may differ from the instruments used to take scrapings.

8. **Fluid Samples:** Many abductees report having blood, saliva, urine, and gynecological fluids taken from them during an examination.

9. **Needle Probes:** Some abductees later find mysterious puncture wounds all over their body, and in sites on their body—such as the calf or forehead—where body fluids would not normally be drawn.

10. **Gynecological Exams:** Some female abductees report being vaginally examined both manually and with instruments during their abduction.

11. **Impregnation:** There are cases in which women seem to have been artificially inseminated with alien sperm during an abduction. The literature chronicles cases in which women begin spotting blood after an abduction and ultrasound tests show an abnormal fetus in early development.

12. **Sexual Relations:** Some male abductees report either having consensual sex or being raped by female alien entities. This may actually be a procedure to have human semen implanted in a female alien as part of a breeding program without artificial means.

13. **Bladder Paralysis:** Some abductees report an inability to urinate for up to forty-eight hours following an abduction medical exam.

14. **Head Surgery:** Some abductees report having had surgery performed inside their heads and many feel that these operations affected (deliberately) their nervous systems in some way.

15. **Eye Damage:** There have been reports of abductees who reported serious eye problems following an abduction experience. It is not known what caused these problems, but it is likely that the alien "doctors" either used a device on the human's eye, radiated it, or instilled some kind of fluid in it that later caused vision problems.

16. **Bone Biopsies:** Some abductees remember having their leg surgically cut open and pieces of their bones removed.

17. **Spinal Taps:** Some abductees report having had lumbar punctures performed on them.

18. **Anesthetizing:** Travis Walton, among others, said he remembered nothing about his abduction experience after a mask was placed over his face.

19. **Implants:** Some abductees report having an implant inserted in one or both of their nostrils. Reportedly, alien implants have been surgically removed by human doctors. Many abductees who experience heavy nosebleeds following an abduction claim to have had something implanted in their nose. Others report having implants inserted into their legs, brains, and elsewhere throughout their body.

20. **Radiation:** Some abductees have found rashes and burns on their bodies that medical authorities have identified as possible radiation burns. Did the aliens expose the abductees to radiation as some form of test? Or were they burned during exposure to the alien craft and/or equipment? The jury's still out.

21. **Limb Pulling:** Some abductees report remembering having their legs and arms roughly tugged on during an examination. One female abductee remembered that only her feet were pulled on by the alien "doctors."

22. **Scanning:** Betty Hill reported an eye-like device scanning her entire body. Abductee Charles Hickson reported being examined by a device identical to the one Betty described.

23. **Endorphin Monitoring:** At least one abductee reported being told that a brain probe performed on him was to monitor his endorphin level. (Endorphins are chemicals that occur naturally in the human brain and have pain-relieving properties

similar to opiates. Endorphins are derived from a substance found in the pituitary gland and are thought to control the activity of the endocrine glands.)

24. **Visual Associations:** Some abductees report being shown intimate home movies of themselves as well as images of Jesus while being monitored and examined.

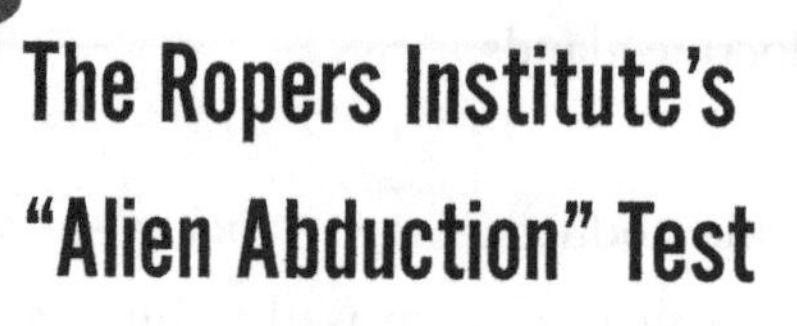

# The Ropers Institute's "Alien Abduction" Test

According to authorities, anyone who can answer "yes" to four of the following five questions has probably been abducted by aliens at some point in his or her life.

The Ropers Institute's findings determined that 2 percent of the six thousand people polled in a 1992 study had been abducted, based on their answers to the following questions. Extrapolating, this means that 5.1 million Americans—approximately one in fifty—have been abducted.

Abduction research from other countries seems to confirm this statistic although many UFOlogists now do not accept the findings of the Ropers poll and several experts have discredited it as erroneous.

1. Have you ever felt paralyzed when you have woken up in the middle of the night, and did you experience being surrounded by strange beings?
2. Have you ever in your life had a "time gap" of more than one hour which you cannot account for?
3. Have you experienced gliding through the air?
4. Have you seen unusual light spheres in your room?
5. Have you had any scars of unknown origin?

# PART VI

# ROSWELL: THE UFO MYSTERY THAT WON'T GO AWAY

# Statements from the Official Air Force Report about the UFO Incident at Roswell

*There can be no doubt that something dropped out of the skies near Roswell on July 4, 1947. The question is what. . . . Everyone agrees that no spaceship wreckage or alien bodies have been made public. Therefore, the truth seeker is left with only human testimony and official pronouncements. The basis for accepting the balloon version rests exclusively on government reports, which deny any unusual aspect to the Roswell case. A lengthy recitation of past official lies, disinformation, and deceit should not be necessary to establish that such pronouncements cannot be accepted at face value.*

—Jim Marrs, *Alien Agenda*

*[I]n Roswell in 1947, the landing of a flying saucer was no fantasy. It was real, the military wasn't able to prevent it, and this time the authorities didn't want a repeat of* War of the Worlds. *So you can see the mentality at work behind the desperate need to keep the story quiet.*

—Philip J. Corso, *The Day After Roswell*

According to countless believers, in 1947, a flying saucer—an extraterrestrial craft—crashed in the desert outside of Roswell, New Mexico.

The legend continues: The wreckage of the saucer was recovered by the United States government, along with the actual bodies of extraterrestrials, the number of which varies, depending on your source. Some of these ETs were dead, at least one was still alive; and they were all transported to Area 51, where our unfortunate visitors, along with the wreckage of their craft, remain to this day.

The Air Force has stated that what crashed at Corona was a Top Secret weather balloon that was part of a program called Project Mogul. It is an understatement to acknowledge that many UFOlogists do not believe the Air Force's "official" version of what happened in the New Mexico desert in 1947.

In 1994, under pressure to finally "come clean," the Air Force released its final statement on the Roswell "incident," as it has come to be known.

This feature looks at twenty-three of the most telling passages from the USAF report, the assertions that pro-Roswell UFOlogists have the hardest time accepting.

You may already be familiar enough with the details of the Roswell crash to find this list interesting and intriguing. For those who want to learn more about what some say happened at Roswell in 1947, there are several books available that go into tremendous detail about the incident. Stanton Friedman's *Crash at Corona* is especially good, as is *The Day after Roswell* by Col. Philip J. Corso (Retired). Reading either or both of these books will likely sway you to the faction that believes that an alien ship *did* crash at Roswell and that the US government has alien corpses in its possession. In the interest of fairness, then, here is what the United States Government says about what happened at Roswell, New Mexico.

> "Secretary of the Air Force Sheila E. Widnall today announced the completion of an Air Force study to locate records that would explain an alleged 1947 UFO incident. Pro-UFO researchers claim an extraterrestrial spacecraft and its alien occupants were recovered near Roswell, N.M., in July 1947 and the fact was kept from the public."

"The Air Force research did not locate or develop any information that the 'Roswell Incident' was a UFO event nor was there any indication of a 'cover-up' by the Air Force. Information obtained through exhaustive records searches and interviews indicated the material recovered near Roswell was consistent with a balloon device of the type used in a then classified project. No records indicated or even hinted at the recovery of 'alien' bodies or extraterrestrial materials."

"Research revealed that the 'Roswell Incident' was not even considered a UFO event until the 1978–1980 time frame. Prior to that, the incident was dismissed because the AAF originally identified the debris recovered as being that of a weather balloon. Subsequently, various authors wrote a number of books claiming that, not only was debris from an alien spacecraft recovered, but also the bodies of the craft's alien occupants. These claims continue to evolve today and the Air Force is now routinely accused of engaging in a 'cover-up' of this supposed event."

"The research located no records at existing Air Force offices that indicated any 'cover-up' by the USAF or any indication of such a recovery. Consequently, efforts were intensified by Air Force researchers at numerous locations where records for the period in question were stored. The records reviewed did not reveal any increase in operations, security, or any other activity in July, 1947, that indicated any such unusual event may have occurred. Records were located and thoroughly explored concerning a then-TOP SECRET balloon project, designed to attempt to monitor Soviet nuclear tests, known as Project Mogul. Additionally, several surviving project personnel were located and interviewed, as was the only surviving person who recovered debris from the original Roswell site in 1947, and the former officer who initially identified the wreckage as a balloon. Comparison of all information developed or obtained indicated that the material recovered near Roswell was consistent with a balloon device and most likely from one of the Mogul balloons that had not been previously recovered. Air Force research efforts did not disclose any records of the recovery of any 'alien' bodies or extraterrestrial materials."

"The modern preoccupation with what ultimately came to be called Unidentified Flying Objects (UFOs) actually began in June, 1947. Although some pro-UFO researchers argue that sightings of UFOs go back to Biblical times, most researchers will not dispute that anything in UFO history can compare with the phenomenon that began in 1947. What was later characterized as 'the UFO Wave of 1940' began with 16 alleged sightings that occurred between May 17 and July 12, 1947 (although some researchers claim there were as many as 800 sightings during that period). Interestingly, the 'Roswell Incident' was not considered one of these 1947 events until the 1978–1980 time frame. There is no dispute, however, that something happened near Roswell in July, 1947, since it was reported in a number of contemporary newspaper articles; the most famous of which were the July 8 and July 9 editions of the *Roswell Daily Record*. The July 8 edition reported 'RAAF Captures Flying Saucer On Ranch In Roswell Region,' while the next day's edition reported, 'Ramey Empties Roswell Saucer' and 'Harassed Rancher Who Located 'Saucer' Sorry He Told About It.'"

"The first story reported that the Intelligence Officer of the 509th Bomb Group, stationed at Roswell AAF, Major Jesse

Marcel, had recovered a 'flying disk' from the rangelands of an unidentified rancher in the vicinity of Roswell and that the disk had been 'flown to higher headquarters.' That same story also reported that a Roswell couple claimed to have seen a large unidentified object fly by their home on July 2, 1947."

"The July 9 edition of the paper noted that Brigadier General Roger Ramey, Commander of the Eighth Air Force at Fort Worth, Texas, stated that upon examination the debris recovered by Marcel was determined to be a weather balloon. The wreckage was described as a '. . . bundle of tinfoil, broken wood beams, and rubber remnants of a balloon. . . .' The additional story of the 'harassed rancher' identified him as W. W. Brazel of Lincoln County, New Mexico. He claimed that he and his son, Vernon, found the material on June 14, 1947, when they 'came upon a large area of bright wreckage made up of rubber strips, tinfoil, a rather tough paper, and sticks.' He picked up some of the debris on July 4 and . . . the next day he first heard about the flying disks and wondered if what he had found might have been the remnants of one of these.' Brazel subsequently went to Roswell on July 7 and contacted the Sheriff, who apparently notified Major Marcel. Major Marcel and 'a man in plain clothes' then accompanied Brazel home to pick up the rest of the pieces. The article further related that Brazel thought that the material: '. . . might have been as large as a table top. The balloon which held it up, if that is how it worked, must have been about 12 feet long, he felt, measuring the distance by the size of the room in which he sat. The rubber was smoky gray in color and scattered over an area about 200 yards in diameter. When the debris was gathered up the tinfoil, paper, tape, and sticks made a bundle about three feet long and 7 or 8 inches thick, while the rubber made a bundle about 18 or 20 inches long and about 8 inches thick. In all, he estimated, the entire lot would have weighed maybe five pounds. There was no sign of any metal in the area which might have been used for an engine and no sign of any propellers of any kind. Although at least one paper fin had been glued onto some of the tinfoil. There were no words to be found anywhere on the instrument although there were letters on some of the parts. Considerable scotch tape and some tape with flowers printed upon it had been used in the construction. No string or wire were to be found but there were some eyelets in the paper to indicate that some sort of attachment may have been used. Brazel said that he had previously found two weather balloons on the ranch, but that what he found this time did not in any way resemble either of these.'"

"There are also now several major variations of the 'Roswell story.' For example, it was originally reported that there was only recovery of debris from one site. This has since grown from a minimal amount of debris recovered from a small area to airplane loads of debris from multiple huge 'debris fields.' Likewise, the relatively simple description of sticks, paper, tape and tinfoil has since grown to exotic metals with hieroglyphics and fiber optic-like materials. Most versions now claim that there were two crash sites where debris was recovered; and at the second site, alleged bodies of extraterrestrial aliens were supposedly retrieved. The number of these 'alien bodies' recovered also varied. These claims are further complicated by the fact that UFO researchers are not in agreement among themselves as to exactly where these recovery sites were located or even the dates of the alleged crash(es)."

"Consistently, however, the AAF was accused of securing these sites, recovering all the material therefrom, keeping locals away, and transporting the recovered wreckage (and bodies) to Roswell under extremely tight security for further processing and later exploitation."

"Once back at Roswell AAF, it is generally alleged that special measures were taken to notify higher headquarters and arrangements made to have recovered materials shipped to other locations for analysis. These locations include Ft. Worth, Texas, the home of the Eighth Air Force Headquarters; possibly Sandia Base (now Kirtland AFB), New Mexico; possibly Andrews AAF, Maryland, and always to Wright Field, now known as Wright Patterson AFB, Ohio. The latter location was the home of 'T-2' which later became known as the Air Technical Intelligence Center (ATIC) and the Air Materiel Command (AMC), and would, in fact, be a logical location to study unknown materials from whatever origin. Most of the Roswell stories that contain the recovery of alien bodies also show them being shipped to Wright Field. Once the material and bodies were dispersed for further analysis and/or exploitation, the government in general, and the Army Air Forces in particular, then engaged in covering up all information relating to the alleged crash and recovery, including the use of security oaths to military persons and the use of coercion (including alleged death threats) to others. This, as theorized by some UFO researchers, has allowed the government to keep the fact that there is intelligent extraterrestrial life from the American public for 47 years. It also supposedly allowed the US Government to exploit recovered extraterrestrial materials by reverse engineering them, ultimately providing such things as fiber optic and stealth technology. The 'death threats,' oaths, and other forms of coercion alleged to have been meted out by the Army Air Forces personnel to keep people from talking have apparently not been very effective, as several hundred people are claimed to have come forward (without harm) with some knowledge of the 'Roswell Incident' during interviews with non-government researchers and the media."

"Adding some measure of credibility to the claims that have arisen since 1978 is the apparent depth of research of some of the authors and the extent of their efforts. Their claims are lessened somewhat, however, by the fact that almost all their information came from verbal reports many years after the alleged incident occurred. Many of the persons interviewed were, in fact, stationed at, or lived near Roswell during the time in question, and a number of them claim military service. Most, however, related their stories in their older years, well after the fact. In other cases, the information provided is second- or third-hand, having been passed through a friend or relative after the principal had died. What is uniquely lacking in the entire exploration and exploitation of the 'Roswell Incident' is official positive documentary or physical evidence of any kind that supports the claims of those who allege that something unusual happened. Conversely, there has never been any previous documentary evidence produced by those who would debunk the incident to show that something did not happen; although logic dictates that bureaucracies do not spend time documenting non-events."

"It was also decided, particularly after a review of the . . . popular literature, that no specific attempt would be made to try to refute, point by point, the numerous claims made in the various publications. Many of these claims appear to be hearsay, undocumented, taken out of context, self-serving, or otherwise dubious."

## WHAT THE "ROSWELL INCIDENT" WAS NOT

"**An Airplane Crash:** Of all the things that are documented and tracked within the Air Force, among the most detailed and scrupulous are airplane crashes. In fact, records of air crashes go back to the first years of military flight. Safety records and reports are available for all crashes that involved serious damage, injury, death, or a combination of these factors. These records also include incidents involving experimental or classified aircraft. USAF records showed that between June 24, 1947, and July 28, 1947, there were five crashes in New Mexico alone, involving A-26C, P-5 IN, C-82A, P-BOA and PQ-14B aircraft; however, none of these were on the date(s) in question nor in the area(s) in question."

"**A Missile Crash:** A crashed or en[emy]-ant[i-aircraft] missile, usually described as a captured German V-2 or one of its variants, is sometimes set forth as a possible explanation for the debris recovered near Roswell. Since much of this testing done at nearby White Sands was secret at the time, it would be logical to assume that the government would handle any missile mishap under tight security, particularly if the mishap occurred on private land. From the records reviewed by the Air Force, however, there was nothing located to suggest that this was the case. Although the bulk of remaining testing records are under the control of the US Army, the subject has also been very well documented over the years within Air Force records. There would be no reason to keep such information classified today. The USAF found no indicators or even hints that a missile was involved in this matter."

"**A Nuclear Accident:** One of the areas considered was that whatever happened near Roswell may have involved nuclear weapons. This was a logical area of concern since the 509th Bomb Group was the only military unit in the world at the time that had access to nuclear weapons. Again, reviews of available records gave no indication that this was the case. A number of records still classified TOP SECRET and SECRET-RESTRICTED DATA having to do with nuclear weapons were located in the Federal Records Center in St. Louis, MO. These records, which pertained to the 509th, had nothing to do with any activities that could have been misinterpreted as the "Roswell Incident." Also, any records of a nuclear-related incident would have been inherited by the Department of Energy (DOE), and, had one occurred, it is likely DOE would have publicly reported it as part of its recent declassification and public release efforts. There were no ancillary records in Air Force files to indicate the potential existence of such records within DOE channels, however."

"**An Extraterrestrial Craft:** The Air Force research found absolutely no indication that what happened near Roswell in 1947, involved any type of extraterrestrial spacecraft. This, of course, is the crux of this entire matter. 'Pro-UFO' persons who obtain a copy of this report, at this point, most probably begin the 'cover-up is still on' claims. Nevertheless, the research indicated absolutely no evidence OF ANY KIND that a spaceship crashed near Roswell or that any alien occupants were recovered therefrom, in some secret military operation or otherwise. This does not mean, however, that the early Air Force was not concerned about UFOs. However, in the early days, 'UFO' meant Unidentified Flying Object, which literally translated as some object in the air that was not readily identifiable. It did not mean, as the term has evolved in today's language, to equate to alien spaceships. Records from the period reviewed by Air Force researchers as well as those cited by the authors mentioned before, do indicate that the USAF MM was seriously concerned about the inability to adequately identify unknown flying objects reported in American airspace. All the records, however, indicated that the focus of concern was not on aliens, hostile or otherwise, but on the Soviet Union. Many documents from that period speak to the possibility of developmental secret Soviet aircraft overlying US airspace. This, of course, was of major concern to the fledgling USAF, whose job it was to protect these same skies."

"The research revealed only one official AAF document that indicated that there was any activity of any type that pertained to UFOs and Roswell in July, 1947. This was a small section of the July Historical Report for the 509th Bomb Group and Roswell AAF that stated: 'The Office of Public Information was quite busy during the month answering inquiries on the 'flying disk,' which was reported to be in possession of the 509th Bomb Group. The object turned out to be a radar tracking balloon.' Additionally, this history showed that the 509th Commander, Colonel Blanchard, went on leave on July 8, 1947, which would be a somewhat unusual maneuver for a person involved in the supposed first ever recovery of extraterrestrial materials. (Detractors claim Blanchard did this as a ploy to elude the press and go to the scene to direct the recovery operations.) The history and the morning reports also showed that the subsequent activities at Roswell during the month were mostly mundane and not indicative of any unusual high-level activity, expenditure of manpower, resources or security."

"Likewise, the researchers found no indication of heightened activity anywhere else in the military hierarchy in the July, 1947, message traffic or orders (to include classified traffic). There were no indications and warnings, notice of alerts, or a higher tempo of operational activity reported that would be logically generated if an alien craft, whose intentions were unknown, entered US territory. To believe that such operational and high-level security activity could be conducted solely by relying on unsecured telecommunications or personal contact without creating any records of such activity certainly stretches the imagination of those who have served in the military who know that paperwork of some kind is necessary to accomplish even emergency, highly classified, or sensitive tasks."

"Similarly, it has also been alleged that General Hoyt Vandenberg, Deputy Chief of Staff at the time, had been involved directing activity regarding events at Roswell. Activity reports located in General Vandenberg's personal papers stored in the Library of Congress did indicate that on July 7, he was busy with a 'flying disk' incident; however this particular incident involved Ellington Field, Texas and the Spokane (Washington) Depot. After much discussion and information gathering on this incident, it was learned to be a hoax. There is no similar mention of his personal interest or involvement in Roswell events except in the newspapers."

"[I]f some event happened that was one of the 'watershed happenings' in human history, the US military certainly reacted in an unconcerned and cavalier manner. In an actual case, the military would have had to order thousands of soldiers and airman, not only at Roswell but throughout the US, to act nonchalantly, pretend to conduct and report business as usual, and generate absolutely no paperwork of a suspicious nature, while simultaneously anticipating that twenty years or more into the future people would have available a comprehensive Freedom of Information Act that would give them great leeway to review and explore government documents. The records indicate that none of this happened (or if it did, it was controlled by a security system so efficient and tight that no one, US or otherwise, has been able to duplicate it since. If such a system had been in effect at the time, it would have also been used to protect our atomic secrets from the Soviets, which history has showed obviously was not the case). The records reviewed confirmed that no such sophisticated and efficient security system existed."

## WHAT THE "ROSWELL INCIDENT" WAS

"As previously discussed, what was originally reported to have been recovered was a balloon of some sort, usually described as a 'weather balloon,' although the majority of the wreckage that was ultimately displayed by General Ramey and Major Marcel in the famous photos in Ft. Worth was that of a radar target normally suspended from balloons. This radar target, discussed in more detail later, was certainly consistent with the description of a July 9 newspaper article which discussed 'tinfoil, paper, tape, and sticks.' Additionally, the description of the 'flying disk' was consistent with a document routinely used by most pro-UFO writers to indicate a conspiracy in progress—the telegram from the Dallas FBI office of July 8, 1947. This document quoted in part states: '. . . The disk is hexagonal in shape and was suspended from a balloon by a cable, which balloon was approximately twenty feet in diameter . . . the object found resembles a high altitude weather balloon with a radar reflector . . . disk and balloon being transported . . .'"

"In addition to those persons above still living who claim to have seen or examined the original material found on the Brazel Ranch, there is one additional person who was universally acknowledged to have been involved in its recovery, Sheridan Cavitt, Lt. Col., USAF (Retired). Cavitt is credited in all claims of having accompanied Major Marcel to the ranch to recover the debris, sometimes along with his Counter Intelligence Corps (CIC) subordinate, William Rickett, who, like Marcel, is deceased. Although there does not appear to be much dispute that Cavitt was involved in the material recovery, other claims about him prevail in the popular literature. He is sometimes portrayed as a closed-mouth (or sometimes even sinister) conspirator who was one of the early individuals who kept the 'secret of Roswell' from getting out. Other things about him have been alleged, including the claim that he wrote a report of the incident at the time that has never surfaced.

Since Lt. Col. Cavitt, who had first-hand knowledge, was still alive, a decision was made to interview him and get a signed sworn statement from him about his version of the events. Prior to the interview, the Secretary of the Air Force provided him with a written authorization and waiver to discuss classified information with the interviewer and release him from any security oath he may have taken. Subsequently, Cavitt was interviewed on May 24, 1994, at his home. Cavitt provided a signed, sworn statement of his recollections in this matter. He also consented to having the interview tape-recorded. In this interview, Cavitt related that he had been contacted on numerous occasions by UFO researchers and had willingly talked with many of them;

however, he felt that he had oftentimes been misrepresented or had his comments taken out of context so that their true meaning was changed. He stated unequivocally, however, that the material he recovered consisted of a reflective sort of material like aluminum foil, and some thin, bamboo-like sticks. He thought at the time, and continued to do so today, that what he found was a weather balloon and has told other private researchers that. He also remembered finding a small 'black box' type of instrument, which he thought at the time was probably a radiosonde. Lt. Col. Cavitt also reviewed the famous Ramey/Marcel photographs of the wreckage taken to Ft. Worth (often claimed by UFO researchers to have been switched and the remnants of a balloon substituted for it) and he identified the materials depicted in those photos as consistent with the materials that he recovered from the ranch. Lt. Col. Cavitt also stated that he had never taken any oath or signed any agreement not to talk about this incident and had never been threatened by anyone in the government because of it. He did not even know the 'incident' was claimed to be anything unusual until he was interviewed in the early 1980."

## CONCLUSION

"The Air Force research did not locate or develop any information that the 'Roswell Incident' was a UFO event. All available official materials, although they do not directly address Roswell per se, indicate that the most likely source of the wreckage recovered from the Brazel Ranch was from one of the Project Mogul balloon trains. Although that project was TOP SECRET at the time, there was also no specific indication found to indicate an official pre-planned cover story was in place to explain an event such as that which ultimately happened. It appears that the identification of the wreckage as being part of a weather balloon device, as reported in the newspapers at the time, was based on the fact that there was no physical difference in the radar targets and the neoprene balloons (other than the numbers and configuration) between Mogul balloons and normal weather balloons. Additionally, it seems that there was over-reaction by Colonel Blanchard and Major Marcel, in originally reporting that a 'flying disk' had been recovered when, at that time, nobody for sure knew what that term even meant since it had only been in use for a couple of weeks.

Likewise, there was no indication in official records from the period that there was heightened military operational or security activity which should have been generated if this was, in fact, the first recovery of materials and/or persons from another world. The post-War US Military (or today's for that matter) did not have the capability to rapidly identify, recover, coordinate, cover-up, and quickly [thwart] public scrutiny of such an event. The claim that they did so without leaving even a little bit of a suspicious paper trail for forty-seven years is incredible.

It should also be noted here that there was little mentioned in this report about the recovery of the so-called 'alien bodies.' This is for several reasons: First, the recovered wreckage was from a Project Mogul balloon. There were no 'alien' passengers therein. Secondly, the pro-UFO groups who espouse the alien bodies theories cannot even agree among themselves as to what, how many, and where, such bodies were supposedly recovered. Additionally, some of these claims have been shown to be hoaxes, even by other UFO researchers. Thirdly, when such claims are made, they are often attributed to people using pseudonyms or who otherwise do not want to be publicly identified, presumably so that some sort of retribution cannot be taken against them (notwithstanding that nobody has been shown to have died, disappeared or otherwise suffered at the hands of the government during the last forty-seven years). Fourth, many of the persons making the biggest claims of 'alien bodies' make their living from the 'Roswell Incident.' While having a commercial interest in something does not automatically make it suspect, it does raise interesting questions related to authenticity. Such persons should be encouraged to present their evidence (not speculation) directly to the government and provide all pertinent details and evidence to support their claims if honest fact-finding is what is wanted. Lastly, persons who have come forward and provided their names and made claims, may have, in good faith but in the 'fog of time,' misinterpreted past events. The review of Air Force records did not locate even one piece of evidence to indicate that the Air Force has had any part in an 'alien' body recovery operation or continuing cover-up."

## Roswell: What We Know for Sure

The standard commentary about the Roswell Incident in July 1947 is that there's nothing new under the sun.

UFO scholars argue that there is overwhelming evidence that a UFO crashed outside of Roswell, attested to at the time by scores of eyewitnesses, while debunkers say the whole story is a myth. Recently a book on Area 51 claimed that Roswell was simply a cover-up to protect the real story of a Soviet plot to crash a Horten Brothers flying wing into the New Mexico desert to frighten local residents and the local Army Air Force base. Even former President Barack Obama has made jokes about Roswell.

But a recent event, the public release of a near-death statement by the 509th's base Public Information Officer, former lieutenant Walter Haut, and his publicly-recorded statements in advance of his sworn statement, give us a much clearer picture of what happened in Roswell during early July 1947.

Haut, who left the Air Force after the services split in August 1947, was the officer who delivered the first public statement to the newspapers that the Army had captured a flying disk outside of Roswell, a statement that made the evening editions of the West Coast newspapers, only to be retracted shortly thereafter when newspapers reported that what had been reported to be a flying saucer was really a weather balloon.

For years, Haut maintained to the public, even after he helped start the

UFO Museum in Roswell, that he knew nothing about a UFO crash and only told the newspapers that he delivered what his commanding officer, Colonel Blanchard, told him to say. Yet just a few years before his death, an elderly Walter Haut began to let slip the story that there was actually a crash of an unidentified object outside of Roswell and that he saw it and knew the entire story. Finally, researchers Don Schmidt and Tom Carey convinced him to make a statement in writing, sworn before a notary public, that he knew the truth about the Roswell incident. The statement, not to be released until after Haut's death so that he would not have broken his promise to Blanchard that he would keep the story secret, was a startling revelation that not only told the whole story of the UFO recovery but the beginning of the official cover-up as well.

Haut wrote that there were two Roswell crash sites, one 40 miles out of Roswell and another 75 miles out of Roswell. The civilian authorities, the fire department, and the county sheriff arrived at one of the crash sites but didn't know about the other. Haut said that he personally visited the UFO debris field and took one of the pieces of the craft, which he described as not from this Earth, back with him as a souvenir for his desk. That piece was later taken from him by an Army CIC officer. Haut said that he saw the craft itself just before it was trucked out of Roswell on a flatbed. He also saw the bodies of the ETs, which Colonel Blanchard indicated to him were about four feet tall.

Perhaps one of the newest bits of information in his statement was his description of the cover-up, which began when Eighth Air Force commander General Rodger Ramey visited the 509th base at Walker Field in Roswell to convene a meeting where he issued orders concerning the cover-up. But even those orders were duplicitous because after saying that the Army would admit to one of the crashes, but not to the other, he ordered base intelligence officer Major Jesse Marcel to fly with the debris to Eighth Air Force headquarters at Fort Worth for a press briefing. However, once at Fort Worth, Ramey switched out the Roswell debris for a weather balloon and ordered

Major Marcel to pose with it for photos while admitting that he has mistakenly identified the weather balloon as a flying saucer, thereby completely undermining Marcel's credibility as an intelligence officer.

None of the debunkers have been able to refute Haut's statement, and thus, for all intents, Haut's sworn eyewitness statement closes the book on Roswell.

# PART VII
# UFOS AND THE BIBLE

# 15 UFO Sightings and Extraterrestrial Visitations Before the Birth of Christ

This chapter could be looked upon as a companion to the chapter "Biblical Passages That Might Describe UFOs or Extraterrestrial Contacts" since it too looks at ancient writings that might have been describing extraterrestrial craft and alien visitations.

As in the biblical writings, these "BC" writings also contain references to unusual lights in the heavens, non-human creatures, strange vessels in the skies, and bizarre sounds from overhead.

Pliny, Cicero, Livy, and the other writers cited here thought they were witnessing manifestations by the Gods.

Maybe we now know better?

1. **BCE 498**

   Cicero, Of the Nature of the Gods, Book I, Chapter 2: "The voices of the Fauns have been heard and deities have appeared in forms so visible that they have compelled everyone who is not senseless or hardened to impiety to confess the presence of the Gods."

2. **BCE 325**

   Livy, History, Book VIII, Chapter 11: "There in the stillness of the night both consuls are said to have been visited by the same apparition, a man of greater than human stature, and more majestic, who declared that the commander of one side and the army of the other must be offered up to the Manes and to Mother Earth."

3. **BCE 223**

Dio Cassius, Roman History, Book I: "At Ariminium a bright light like the day blazed out at night; in many portions of Italy three moons became visible in the night time."

4. **BCE 222**

Pliny, Natural History, Book II, Chapter 32: "Also three moons have appeared at once, for instance, in the consulship of Gnaeus Domitius and Gaius Fannius."

5. **BCE 218**

Livy, History, Books XXI-XXII: "In Amiterno district in many places were seen the appearance of men in white garments from far away. The robe of the sun grew smaller. At Praeneste glowing lamps from heaven. At Arpi a shield in the sky. The moon contended with the sun and during the night two suns were seen. Phantom ships appeared in the sky."

6. **BCE 217**

Livy, History, Book XXII, Chapter 1: "At Falerii the sky had seemed to be rent as it were with a great fissure and through the opening a bright light had shone."

7. **BCE 214**

Julius Obsequens, Prodigiorum Libellus, Chapter 66: "At Hadria an altar was seen in the sky and about it the forms of men in white clothes."

**8. BCE 163**

Julius Obsequens, Prodigiorum Libellus: "In the consulship of Tiberius Gracchus and Manius Juventus at Capua the sun was seen by night. At Formice two suns were seen by day. The sky was afire. In Cephallenia a trumpet seemed to sound from the sky. There was a rain of earth. A windstorm demolished houses and laid crops flat in the field. By night an apparent sun shone at Pisaurum."

**9. BCE 122**

Julius Obsequens, Prodigiorum Libellus, Chapter 114: "In Gaul three suns and three moons were seen."

**10. BCE 91**

Julius Obsequens, Prodigiorum Libellus, Chapter 114: "Near Spoletium a gold-colored fireball rolled down to the ground, increased in size, seemed to move off the ground toward the east and was big enough to blot out the sun."

**11. BCE 85**

Pliny, Natural History, Book II, Chapter 34: "In the consulship of Lucius Valerius and Caius Marius a burning shield scattering sparks ran across the sky."

**12. BCE 66**

Pliny, Natural History, Book II, Chapter 35: "In the consulship of Gnaeus Octavius and Gaius Suetonius a spark was seen to fall from a star and increase in size as it approached the earth. After becoming as large as the moon it diffused a sort of cloudy daylight and then returning to the sky changed into a torch. This is the only record of its occurrence. It was seen by the pro-counsel Silenus and his suite."

**13. BCE 48**

Dio Cassius, Roman History, Book IV: "Thunderbolts had fallen upon Pompey's camp. A fire had appeared in the air over Caesar's camp and had fallen upon Pompey's . . . In Syria two young men announced the result of the battle and vanished."

**14. BCE 42**

Julius Obsequens, Prodigiorum Libellus, Chapter 130: "In Rome light shone so brightly at nightfall that people got up to begin work as though day had dawned. At Murtino three suns were seen about the third hour of the day, which presently drew together in a single orb."

**15. BCE ??**

Cicero, On Divination, Book I, Chapter 43: "How often has our Senate enjoined the decemvirs to consult the books of the Sibyl! For instance, when two suns had been seen or when three moons had appeared and when flames of fire were noticed in the sky; or on that other occasion when the sun was beheld in the night, when noises were heard in the sky, and the heaven itself seemed to burst open, and strange globes were remarked in it."

# 40 Biblical Passages That Might Describe UFOs or Extraterrestrial Contacts

Many UFOlogists and UFO buffs point to passages in the Bible (and other sacred writings) that can easily be "reinterpreted" within a "UFO/alien visitation" context if one so desires.

The contention is that any references to voices from heaven, figures of light, angels from "heaven," fire from "heaven," moving stars, figures descending from heaven, pillars of smoke, etc., are actually describing alien spacecraft and extraterrestrial visitations—but the ancients were incapable of adequately chronicling these events using anything but metaphor and symbolism.

Here is a look at forty of the most commonly cited "UFO" biblical passages. For more on this subject, see the chapter, "Was Jesus an Extraterrestrial?" Theory. (And, yes, the "wheel of fire" passage from Ezekiel and the "chariot of fire" passage from II Kings are both included.)

**NOTE:** All biblical citations are from the King James version of the Old Testament and the New Testament.

1. **Exodus 3:2:** And the angel of the Lord appeared unto him in a flame of fire out of the midst of a bush, and he looked, and, behold, the bush burned with fire, and the bush was not consumed.

2. **Exodus 13:21-22:** And the Lord went before them by day in a pillar of a cloud, to lead them the way, and by night in a pillar of fire, to give them light; to go by day and night: He took not away the pillar of the cloud by day, nor the pillar of fire by night, from before the people.

3. **Exodus 16:10:** And it came to pass, as Aaron spake unto the whole Congregation of the children of Israel, that they looked toward the wilderness, and, behold, the glory of the Lord appeared in the cloud.

4. **Exodus 19:9:** And the Lord said unto Moses, Lo, I come unto thee in a thick cloud, that the people may hear when I speak with thee, and believe thee for ever. And Moses told the words of the people unto the Lord.

5. **Exodus 19:16:** And it came to pass on the third day in the morning, that there were thunders and lightnings, and a thick cloud upon the mount, and the voice of the trumpet exceeding loud; so that all the people that was in the camp trembled.

6. **Exodus 19:18:** And mount Sinai was altogether on a smoke, because the Lord descended upon it in fire: and the smoke thereof ascended as the smoke of a furnace, and the whole mount quaked greatly.

7. **Exodus 24:15-18:** And Moses went up into the mount, and a cloud covered the mount.

And the glory of the Lord abode upon mount Sinai, and the cloud covered it six days: and the seventh day he called unto Moses out of the midst of the cloud.

And the sight of the glory of the Lord was like devouring fire on the top of the mount in the eyes of the children of Israel.

And Moses went into the midst of the cloud, and gat him up into the mount: and Moses was in the mount forty days and forty nights.

8. **Exodus 33:9-10:** And it came to pass, as Moses entered into the tabernacle, the cloudy pillar descended, and stood at the door of the tabernacle, and the Lord talked with Moses.

   And all the people saw the cloudy pillar stand at the tabernacle door: and all the people rose up and worshipped, every man in his tent door.

9. **Exodus 34:5:** And the Lord descended in the cloud, and stood with him there, and proclaimed the name of the Lord.

10. **Exodus 34:29-30:** And it came to pass, when Moses came down from mount Sinai with the two tables of testimony in Moses' hand, when he came down from the mount, that Moses wist not that the skin of his face shone while he talked with him.

    And when Aaron and all the children of Israel saw Moses, behold, the skin of his face shone; and they were afraid to come nigh him.

11. **Leviticus 16:2:** And the Lord said unto Moses, Speak unto Aaron thy brother, that he come not at all times into the holy place within the vail before the mercy seat, which is upon the ark; that he die not: for I will appear in the cloud upon the mercy seat.

12. **Numbers 11:1:** And when the people complained, it displeased the Lord and the Lord heard it; and his anger was kindled;

and the fire of the Lord burnt among them, and consumed them that were in the uttermost parts of the camp.

13. **Numbers 12:5-6:** And the Lord came down in the pillar of the cloud, and stood in the door of the tabernacle, and called Aaron and Miriam: and they both came forth.

    And he said, Hear now my words: If there be a prophet among you, I the Lord will make myself known unto him in a vision, and will speak unto him in a dream.

14. **Numbers 12:9-10:** And the anger of the Lord was kindled against them; and he departed.

    And the cloud departed from off the tabernacle; and, behold, Miriam became leprous, white as snow: and Aaron looked upon Miriam, and, behold, she was leprous.

15. **Deuteronomy 31:15:** And the Lord appeared in the tabernacle in a pillar of a cloud: And the pillar of the cloud stood over the door of the tabernacle.

16. **Judges 20: 40:** But when the flame began to arise up out of the city with a pillar of smoke, the Benjamites looked behind them, and, behold, the flame of the city ascended up to heaven.

17. **I Kings 18:38:** Then the fire of the Lord fell, and consumed the burnt sacrifice, and the wood, and the stones, and the dust, and licked up the water that was in the trench.

18. **I Kings 19:11-13:** And he said, Go forth, and stand upon the mount before the Lord. And, behold, the Lord passed by, and a great and strong wind rent the mountains, and brake

in pieces the rocks before the Lord; but the Lord was not in the wind: and after the wind an earthquake; but the Lord was not in the earthquake:

And after the earthquake a fire; but the Lord was not in the fire: and after the fire a still small voice.

19. **II Kings 2:1:** And it came to pass, when the Lord would take up Elijah into heaven by a whirlwind, that Elijah went with Elisha from Gilgal.

20. **II Kings 2:8:** And Elijah took his mantle, and wrapped it together, and smote the waters, and they were divided hither and thither, so that they two went over on dry ground.

21. **II Kings 2:11:** And it came to pass, as they still went on, and talked, that, behold, there appeared a chariot of fire, and horses of fire, and parted them both asunder; and Elijah went up by a whirlwind into heaven.

22. **II Chronicles 7:1:** Now when Solomon had made an end of praying, the fire came down from heaven, and consumed the burnt offering and the sacrifices; and the glory of the Lord filled the house.

23. **Nehemaiah 9:11-12:** And thou didst divide the sea before them, so that they went through the midst of the sea on the dry land; and their persecutors thou threwest into the deeps, as a stone into the mighty waters.

    Moreover thou leddest them in the day by a cloudy pillar; and in the night by a pillar of fire, to give them light in the way wherein they should go.

24. **Psalms 18:8:** There went up a smoke out of his nostrils, and fire out of his Mouth devoured: coals kindled by it.

25. **Psalms 18:10:** And he rode upon a cherub, and did fly: yea, he did fly upon the wings of the wind.

26. **Psalm 68:4:** Sing unto God, sing praises to his name, extol him that rideth upon the heavens by his name, JAH, and rejoice before him.

27. **Psalms 99:7:** He spake unto them in the cloudy pillar: they kept his testimonies, and the ordinance that he gave them.

28. **Isaiah 6:1-4:** In the year that king Uzziah died I saw also the Lord sitting upon a throne, high and lifted up, and his train filled the temple.

    Above it stood the seraphims: each one had six wings; with twain he covered and with twain he did fly.

    And one cried unto another, and said, Holy, holy, holy, is the Lord of hosts: the whole earth is full of his glory.

    And the posts of the door moved at the voice of him that cried, and the house was filled with smoke.

29. **Ezekiel 1:1-28:** Now it came to pass in the thirtieth year, in the fourth month, in the fifth day of the month, as I was among the captives by the river of Chebar, that the heavens were opened, and I saw visions of God.

    In the fifth day of the month, which was the fifth year of king Jehoiachin's captivity,

    The word of the Lord came expressly unto Ezekiel the priest, the son of Buzi, in the land of the Chaldeans by the river Chebar; and the hand of the Lord was there upon him

And I looked, and, behold, a whirlwind came out of the north, a great cloud, and a fire infolding itself, and a brightness as about it, and out of the midst thereof as the colour of amber, out of the midst of the fire.

Also out of the midst thereof came the likeness of four living creatures. And this was their appearance; they had the likeness of a man.

And every one had four faces, and every one had four wings.

And their feet were straight feet; and the sole of their feet as like the sole of a calf's foot: and they sparkled like the colour of burnished brass.

And they had the hands of a man under their wings on their four sides; and they four had their faces and their wings.

Their wings were joined one to another; they turned not when they went; they went every one straight forward.

As for the likeness of their faces, they four had the face of a man, and the face of a lion, on the right side: and they four had the face of an ox on the left side; they four also had the face of an eagle.

Thus were their faces: and their wings were stretched upward; two wings of every one were joined one to another, and two covered their bodies.

And they went every one straight forward: whither the spirit was to go, they went; and they turned not when they went.

As for the likeness of the living creatures, their appearance was like burning coals of fire, and like the appearance of lamps: it went up and down among the living creatures; and the fire was bright, and out of the fire went forth lightning.

And the living creatures ran and returned as the appearance of a flash of lightning.

Now as I beheld the living creatures, behold one wheel upon the earth by the living creatures, with his four faces.

The appearance of the wheels and their work was like unto the colour of a beryl: and they four had one likeness: and their appearance and their work was as it were a wheel in the middle of a wheel.

When they went, they went upon their four sides: and they turned not when they went.

As for their rings, they were so high that they were dreadful; and their rings were full of eyes round about them four.

And when the living creatures went, the wheels went by them: and when the living creatures were lifted up from the earth, the wheels were lifted up.

Whithersoever the spirit was to go, they went, thither was their spirit to go; and the wheels were lifted up over against them: for the spirit of the living creature was in the wheels.

When those went, these went; and when those stood, these stood; and when those were lifted up from the earth, the wheels were lifted up over against them: for the spirit of the living creature was in the wheels.

And the likeness of the firmament upon the heads of the living creature was as the colour of the terrible crystal, stretched forth over their heads above.

And under the firmament were their wings straight, the one toward the other: every one had two, which covered on this side, and every one had two, which covered on that side, their bodies.

And when they went, I heard the noise of their wings, like the noise of great waters, as the voice of the Almighty, the voice of speech, as the noise of an host: when they stood, they let down their wings.

And there was a voice from the firmament that was over their heads, when they stood, and had let down their wings.

And above the firmament that was over their heads was the likeness of a throne, as the appearance of a sapphire stone: and upon the likeness of the throne was the likeness as the appearance of a man above upon it.

And I saw as the colour of amber, as the appearance of fire round about within it, from the appearance of his loins even upward, and from the appearance of his loins even downward, I saw as it were the appearance of fire, and it had brightness round about.

As the appearance of the bow that is in the cloud in the day of rain, so was the appearance of the brightness round about. This was the appearance of the likeness of the glory of the Lord. And when I saw it, I fell upon my face, and I heard a voice of one that spake.

30. **Matthew: 17:1-8:** And after six days Jesus taketh Peter, James, and John his brother, and bringeth them up into an high mountain apart,

And was transfigured before them: and his face did shine as the sun, and his raiment was white as the light.

And, behold, there appeared unto them Moses and Elias talking with him.

Then answered Peter, and said unto Jesus, Lord, it is good for us to be here: if thou wilt, let us make here three tabernacles; one for thee, and one for Moses, and one for Elias.

While he yet spake, behold, a bright cloud overshadowed them: and behold a voice out of the cloud, which said, This is my beloved Son, in whom I am well pleased; hear ye him.

And when the disciples heard it, they fell on their face, and were sore afraid.

And Jesus came and touched them, and said, Arise, and be not afraid.

And when they had lifted up their eyes, they saw no man, save Jesus only.

31. **Acts 1:9-11:** And when he had spoken these things, while they beheld, he was taken up; and a cloud received him out of their sight.

    And while they looked stedfastly toward heaven as he went up, behold, two men stood by them in white apparel;

    Which also said, Ye men of Galilee, why stand ye gazing up into heaven? this same Jesus, which is taken up from you into heaven, shall so come in like manner as ye have seen him go into heaven.

32. **Acts 7:30:** And when forty years were expired, there appeared to him in the wilderness of mount Sina an angel of the Lord in a flame of fire in a bush.

33. **Acts 7:35:** This Moses whom they refused, saying, Who made thee a ruler and a judge? The same did God send to be a ruler and a deliverer by the hand of the angel which appeared to him in the bush.

34. **Acts 7:38:** This is he, that was in the church in the wilderness with the angel which spake to him in the mount Sina, and with our fathers: who received the lively oracles to give unto us.

35. **Acts 7:52-53:** Which of the prophets have not your fathers persecuted? And they have slain them which shewed before of the coming of the Just One; of whom ye have been now the betrayers and murderers:

Who have received the law by the disposition of angels, and have not kept it.

36. **II Peter 1:17-18:** For he received from God the Father honour and glory, when there came such a voice to him from the excellent glory, This is my beloved Son, in whom I am well pleased.

    And this voice which came from heaven we heard, when we were with him in the holy mount.

37. **Luke 1:26:** And in the sixth month the angel Gabriel was sent from God unto a city of Galilee, named Nazareth.

38. **Matthew 2:9-10:** When they had heard the king, they departed; and, lo, the star, which they saw in the east, went before them, till it came and stood over where the young child was.

    When they saw the star, they rejoiced with exceeding great joy.

39. **Luke 3:22:** And the Holy Ghost descended in a bodily shape like a dove upon him, and a voice came from heaven, which said, Thou art my beloved Son; in thee I am well pleased.

40. **Matthew 28:2-4:** And, behold, there was a great earthquake: for the angel of the Lord descended from heaven, and came and rolled back the stone from the door, and sat upon it.

    His countenance was like lightning, and his raiment white as snow: And for fear of him the keepers did shake, and became as dead men.

# 6 Similarities Between Apparitions of the Virgin Mary and UFO Sightings

*Apparitions of the Virgin Mary would seem to be a logical form of communication initiated by ETIs [extraterrestrial intelligences] and directed to us.*

—from "The Guadalupe Event"
by Johannes Fiebag, Ph.D

The Blessed Virgin Mary has appeared to many of the Roman Catholic faithful for centuries. The devout believe that Mary makes herself visible to impart a message of hope or a warning, and that she allows herself to be seen only by a select few.

Fatima and Medjugore are two of the most famous Marian apparitions, and yet Mary is said to appear all over the world, including in the United States, on a fairly regular basis.

UFOlogists, however, are not as willing to accept sightings of the Virgin Mary as divine manifestations and, in fact, many believe that what the faithful are actually seeing are UFOs and alien beings. These debunkers point to astonishing similarities between reported UFO sightings and apparitions of Mary and suggest that perhaps these visions are not religious in nature, but extraterrestrial.

Here are some of the most commonly cited similarities between sightings of the Virgin Mary and sightings of UFOs.

1. **Luminous spheres:** Many sightings of the Virgin Mary include descriptions of luminous spheres—globes of ethereal light. These spheres are also a common element of UFO sightings.

2. **The "falling leaf" motion:** Visions of the Virgin Mary often include a description of lights surrounding her and descending in a swaying motion known in UFOlogy as the "falling leaf motion." (The lights sway from side to side as they slowly descend from the sky.)

3. **Heat waves:** Many UFO witnesses have reported feeling waves of heat wash over them. Similar accounts of heat waves have been a common element of sightings of the Virgin Mary.

4. **Angel hair:** Angel hair consists of thin, cotton-like fibers that are found at the scene of UFO sightings. The strands look like spider webs (which many skeptics assert is precisely what they are) and have been found on houses, on the ground, in trees, and hanging off electrical wires following a UFO sighting. Angel hair has likewise been reported at sightings of the Virgin Mary, one of the most notable being the May 1917 vision of Mary in Fatima, Portugal.

5. **Messages:** UFO sighters often report having been given a message from the occupants of the craft, either telepathically or sometimes visually. People who claim to have seen the Virgin Mary often also report being given messages. Again, the appearance of Mary at Fatima is an example of the Virgin giving messages to those who can see her.

6. **Healings:** Some who have had close encounters have reported miraculous healings after their encounters with a UFO.

> Similar unexplainable healings also often accompany apparitions of the Virgin Mary and many of the faithful pray to Mary for a miracle in apparently hopeless medical situations. Many claim to have had their prayers answered and to have been healed by the Virgin.

Also, some UFOlogists believe that the *figure* of the Virgin Mary so often seen during apparitions may be a holographic projection by extraterrestrial intelligences and that these projections may be accompanied and enhanced by some kind of mass hypnosis induced by either unknown chemical agents or lights that affect brain waves.

# Was Jesus an Extraterrestrial?

*A city that is set on a hill cannot be hid.*

—Jesus Christ, in The Gospel According to Matthew, Chapter 5, Verse 14

## Was Jesus Christ an extraterrestrial?

There are people who believe so, especially Dr. Barry Downing, who wrote an entire book (*The Bible and Flying Saucers*) about the how Biblical imagery and metaphor was, in fact, describing alien visitations and UFO sightings and landings.

Downing, among others, believes that Jesus was an alien sent to Earth as an undercover scout, and that it was probably not the alien's intention for an entire religion to be created around his presence.

Here is a look at some of the most compelling "ET" interpretations of the idea that Jesus was an extraterrestrial.

**Jesus was the Son of God:** Jesus was born of a virgin.
**Jesus was an ET:** Jesus was a genetically engineered being who was implanted in an Earth woman to be born on Earth.

**Son of God:** The angel Gabriel visited Mary to tell her that she would be conceived by the Holy Spirit.
**ET:** An alien emissary visited Mary to inform her of the implantation.

**Son of God:** The star of Bethlehem guided the Three Wise Men to the place of Jesus's birth.
**ET:** An alien ship flew across the sky within plain view of three alien scouts whose assignment it was to monitor the alien/human birth.

**Son of God:** In *The Gospel According to Matthew*, Chapter 3, Verses 16 and 17, Jesus is baptized, and Matthew writes, "And Jesus, when he was baptized, went up straightaway out of the water, and, lo, the heavens were opened unto him, and he saw the Spirit of God descending like a dove, and lighting upon him. And lo a voice from heaven, saying, 'This is my beloved Son, in whom I am well pleased."
**ET:** Jesus the ET was teleported out of the water by an anti-gravity ray beamed down upon him from an alien spacecraft, and then something was said to him that was unintelligible to the onlookers, but which they interpreted to mean what they wanted it to mean anyway.

**Son of God:** Jesus walked on water.
**ET:** Jesus the ET had an anti-gravity device that allowed him to appear as though he was walking on water.

**Son of God:** Jesus raised Lazarus from the dead.
**ET:** Jesus the ET had advanced medical abilities that allowed him to revive Lazarus, who was probably not dead but just in a deep coma.

**Son of God:** In *The Gospel According to John*, Chapter 8, Verse 23, Jesus tells his followers, "You are from beneath; I am from above. Ye are of this world; I am not of this world."
**ET:** "You are Earthlings; I am an extraterrestrial."

**Son of God:** Jesus changed water into wine.
**ET:** Jesus the ET had a device that could transmute at the subatomic level the molecules of water into wine molecules.

**Son of God:** Jesus miraculously multiplied the loaves and fishes.
**ET:** Jesus the ET had a technologically advanced food synthesizing device that "read" the original and was then able to duplicate as many completely real duplicates as desired.

**Son of God:** Jesus spent forty days and forty nights in the desert.
**ET:** Jesus spent forty days and forty nights on the Mothership reporting on his terrestrial findings and taking a much-needed rest.

**Son of God:** Jesus rose from the dead.
**ET:** At the moment when onlookers saw Jesus "commend his spirit" to his heavenly Father, his alien colleagues flying above him in the Mothership put him into suspended animation and then "reactivated" him when he was lying in his tomb.

**Son of God:** Jesus physically ascended into Heaven.
**ET:** The ET known as Jesus was "beamed up" to the mothership as his followers watched. (This is also how his mother Mary "ascended into Heaven.")

**Son of God:** Jesus spoke to Paul as Paul traveled the road to Damascus and Paul was instantly converted to Christianity.

**ET:** The ET Jesus spoke to Paul from a spaceship, so terrifying and awing him that he willingly agreed to do whatever the voice told him to.

**Son of God:** In *1 Thessalonians*, Chapter 4, Verses 16 and 17, Paul writes, "For the Lord himself shall descend from heaven with a shout, with the voice of the archangel, and with the trumpet of God, and the dead in Christ shall rise first. Then we which are alive and remain shall be caught up together with them in the clouds, to meet the Lord in the air, and so shall we ever be with the Lord."

**ET:** When Jesus the ET returns to Earth for his followers, he will first raise the dead who believed in him and beam them up to the Mothership, followed by those living followers of his who will also be beamed up to the "clouds," where the Mothership awaits them all.

PART VIII

# AS WEIRD AS IT GETS

# All the Presidents' UFOs

UFOs and presidential disclosures of UFOs have been a part of American history from the days of General George Washington at Valley Forge—where he saw a floating orb hover before his eyes and little figures climb out of it—through Presidents Jimmy Carter's and Ronald Reagan's sightings, to Democratic candidate Hillary Clinton's promise to release all of our nation's UFO files that do not compromise national security.

The calls for official presidential disclosure of the UFO phenomenon have resounded through the years, but few of the pundits calling for disclosure mention that it was President Harry Truman in 1950 who told the American people that what they had seen in the skies they couldn't identify were UFOs. "These are flying saucers," President Truman told the press. "And we know what they are." It was a stunning admission of the presence of strange craft in our skies, which came a full two years before the UFO invasion of Washington, DC, in 1952, an event captured on motion picture newsreel cameras as well as in still photos on the covers of our nation's newspapers.

After George Washington's disclosure of a UFO sighting to his staff at Valley Forge, our nation's third president, Thomas Jefferson, revealed to the American Philosophical Society at their meeting in Philadelphia that he had been told of a UFO sighting in Baton Rouge—a craft that, in Jefferson's words "was as large as a house"—and that the event was multiply witnessed by citizens who were sober, serious, and not given to flights of fancy.

For the ensuing century, through the presidencies of Buchanan, Pierce, and Lincoln, there were reports of many paranormal events, but the next clear UFO sighting, reported in newspapers, was the sighting of a UFO over President Theodore Roosevelt's house at Sagamore Hill. Teddy Roosevelt was well aware of this sighting and, UFO historians argue, he was extremely impressed with it. As a precursor to what Ronald Reagan told Soviet Chairman Gorbachev and then the United Nations General Assembly, Roosevelt argued for an international body to bring nations together to forge a codification of the rules of war and an international peace treaty.

Another Roosevelt, FDR, also had his presidency impacted by UFOs, this time during World War II when intelligently guided balls of light, called "Foo Fighters," intercepted American fighters escorting B-17 bombers over occupied Europe and Germany. The Americans thought these balls of light that seemed to hover around their planes and even move with them were Nazi secret weapons. But the Luftwaffe pilots, who also saw them, believed they were Allied secret weapons. And Foo Fighters didn't just appear over Europe. They also appeared over the South Pacific where one of them was shot down by a gunner on a B-29 heavy bomber. The Foo Fighter blew up into fragments and disappeared.

President Harry Truman could be called our nation's first UFO president because it was on his watch in 1947 that the purported crash of a UFO at Roswell took place. But instead of simply being a witness to reports of the crash, President Truman took affirmative action and ordered the Army to find the lowest common denominator of technology retrieved from the crash and deliver it to American companies working on similar technologies.

The first piece of debris to reach American industry was a piece of electronic circuitry from the crashed craft, which was sent to Bell Labs where scientists were working on an electronic switch to replace the Edison tube, a glass-enclosed vacuum tube with a filament running from pole to pole, essentially an incandescent light bulb. But scientists Brittain and Shockley had failed to develop a device that could act as a transmitter of electrons as well as a resistor that could release one electron at a time across the circuit.

When the scientists received the piece of circuitry from the crashed UFO, they were able to discern its chemical composition and reverse engineer a device built on a silicon base doped with arsenic that allowed one electron at a time to pass across it. This combination of a transmitter and resistor was named a transistor, which Bell Labs patented in 1948 and won its commercial patent the following year. By the early 1950s, tiny transistor radios became some of the most popular devices of the decade, and the transistor went on to become the main component of electronic circuit boards, spurring the evolution of digital computers and today's Internet Age.

President Eisenhower, who succeeded Truman in office, was also an eyewitness to UFOs when they appeared over the North Sea during the first NATO joint naval, air, and ground exercises designed to prevent the Soviet Union from sending its submarines out of the North Sea into the Atlantic. Witnesses aboard the bridge of the new US aircraft carrier, the USS *FDR*, witnessed UFOs flying out of the water and shooting off into the atmosphere, events also witnessed by then-retired General Eisenhower, who cautioned the bridge crew not to talk about what they saw. In July 1952, Ike had learned about the invasion of UFOs over the restricted Washington, DC, airspace and was then briefed when he took office in 1953 about the crash at Roswell and what the military believed they learned from the analysis of that craft.

There is a story long believed by UFO historians that President Eisenhower was asked to meet with extraterrestrials at MUROC Air Force Base where the president was said to have structured a form of an "open skies" treaty with them. The basic premise of this treaty was that we would not interfere with their presence in our skies, allowing them to surveil us, in return for which the ETs would provide us with advanced technology. While no records of this meeting or this treaty have come to light, Eisenhower's structuring the same type of agreement with the Soviet Union—called "Open Skies"—to prevent each side's launching a nuclear attack by mistake, leads people to believe that the UFO treaty was the basis for this agreement with the Soviet Union.

Many UFO historians and conspiracy theorists believe that it was the UFO question that resulted in the assassination of President John F. Kennedy. During his romantic relationship with movie star Marilyn Monroe, Kennedy was said to have engaged in pillow talk during which he revealed the secrets not only of Area 51 but talked about what Marilyn said were "little men from space" that the US kept in captivity there. When she threatened Attorney General Robert F. Kennedy over the phone with exposing her affair with both him and his brother, she referred to what the president told her about crashed debris from space and "little people," a conversation that was wiretapped and transcribed by the CIA. If JFK had been revealing some of our nation's most closely guarded UFO secrets, the CIA knew about it and took steps to keep further leaks from taking place. The year was 1962. Marilyn died that year, apparently of suicide, although questions surrounding her death remain unanswered, and JFK was assassinated the following year.

UFOs also played a role during President Lyndon Johnson's administration, not only flying over US installations in Vietnam during the Vietnamese War but also at the crash in Kecksburg, Pennsylvania in 1965, where the army guarded the crash site while a crew from NASA retrieved the craft. The Johnson administration was also made aware of the 1966 Hillsdale, Michigan, UFO, which prompted then Representative and later President Gerald Ford to write to Mendel Rivers of the House Armed Services Committee to open hearings on UFOs.

Lyndon Johnson was followed in office by President Richard Nixon, who, according to the late television and film star Jackie Gleason, allowed Gleason to travel to Homestead Air Force Base in Florida where he viewed the body of an extraterrestrial. Although Gleason had been badgering Nixon for years to tell him about the presence of UFOs, Nixon had refused him. Finally, Nixon relented and arranged for Gleason to see the proof of UFOs for himself, an event that so unnerved Gleason that he told his wife and his costars at Sony Pictures where he was filming *Nothing in Common* with Tom Hanks.

Just as a too-aggressive policy on UFO disclosure might have contributed to JFK's assassination, so did it almost contribute to President Gerald

Ford's after he told people after he ascended to the Oval Office upon Nixon's resignation that he would reveal the truth about UFOs to the American people. When he was in California, not one, but two members of the notorious Manson family made assassination attempts on Ford. Was it because of his UFO promise? Charles Manson himself revealed to psychologist and author Joel Norris (*Serial Killers: A Growing Menace*, New York: Doubleday, 1988) that he was working for the US government, specifically, the CIA under nonofficial cover. Were the assassination attempts on Ford retribution for Ford's promise to reveal the truth about UFOs? UFO historians hold that as a common belief.

Gerald Ford was succeeded in office by Jimmy Carter, who defeated him for reelection in 1976. Carter had experienced a UFO sighting himself in Georgia years earlier, a sighting that he described to a UFO reporting agency in a formal report. [See Carter's report in this volume.] Carter held to his belief that what he saw was not the moon, not Venus, nor the north star, nor an airplane. And the fact that Carter was an experienced Naval officer, a submariner, made his sighting all the more credible. When asked on the campaign trail whether he would reveal the truth about UFOs to the American people, Carter said that he would. He went back on his promise after one of his advisors, one with top security clearance, entered the Oval Office with a threat that if he talked about UFOs, his presidency would end and his family would be endangered.

Carter, as president-elect, also asked his Director of Intelligence during his transition briefing if he would be told what the CIA knew about UFOs. But the director, George H. W. Bush, told him flatly, "Mr. President-Elect, you have no need to know."

Jimmy Carter was succeeded in office by California governor Ronald Reagan, who, as governor, had two separate UFO sightings, one over the Mojave desert and another over the Pacific Coast Highway near Malibu. The first sighting took place as Reagan was flying in his plane from Los Angeles to Sacramento, the state capitol. He spotted a UFO through the window and directed his pilot to follow it over the desert. The UFO simply disappeared from view at a fantastic speed. Reagan was so impressed with what he saw, when he landed he told the Sacramento Bureau Chief for the *Wall Street Journal* exactly what he saw and that he would tell Nancy as soon as he got to the Governor's Mansion.

Reagan's second sighting occurred when a UFO flew over the Pacific Coast Highway as the governor and his first lady were driving to a surprise party in Hollywood for the late Hollywood film star, William Holden (*Network, Love is a Many Splendored Thing*). Reagan beguiled the guests at the party with his UFO story, which so impressed television personality Lucille Ball (*I Love Lucy*) that she told all of her friends.

Ronald Reagan's UFO experiences, historians say, prompted his idea of the Space Defense Initiative, or the "Star Wars" weapon system, not to shoot down Soviet missiles but, more importantly, to defend the planet against UFOs.

Reagan's vice president was George H. W. Bush, whom he had defeated in the 1980 Republican presidential primary, and who had told President Jimmy Carter that he had no need to know what the CIA had in its files on UFOs. He repeated that same statement years later during his son Jeb's primary campaign, telling people who asked him about UFOs that Americans "can't handle the truth about UFOs."

Perhaps some of the most intriguing questions about how UFOs play into national security came out of the Clinton administration that began in 1991 when Bill Clinton asked his longtime friend and incoming Assistant Attorney General Webster Hubbell to investigate the crash at Roswell as well as the Kennedy assassination. Hubbell, Clinton said, found nothing except for the chilling fact that there was "a secret government within the government," above the president's top security clearance.

Even more chilling, though, was the event that took place on March 13, 1997, known as the Phoenix Lights—appearances of a formation of lights floating southeast from Henderson, Nevada, across Arizona, to the Mexico border. This was also the night of the appearance of the Hale-Bopp comet and the mass suicide of the Heaven's Gate cult. That particular night, President Clinton was staying at the home of golfer Greg Norman, when, inexplicably at first, Clinton went incommunicado. It was said that the president had injured his knee. However, on that night, formations of lights flew across Phoenix at 8:30 p.m. and then again at 10:30 p.m. After an initial refusal to talk about the lights, the Air Force eventually said that they were the result of flares dropped during an Air National Guard training exercise by a squadron from Maryland. That might have explained the 10:30 lights to some, but not the 8:30 lights. What made the 8:30 lights even more tantalizing was that they were not only witnessed by scores of residents in Paradise Valley, Arizona, but witnesses also saw a structured craft, a giant flying triangle, with the lights on the three corners of the craft.

The frenzy among panicked Phoenix residents over the event, which was heavily covered in the national media, was so great that Arizona governor Fife Symington (whose cousin Stuart Symington had been the first Secretary of the Air Force after the services split in August 1947—a month after Roswell) held a news conference. Symington said the conference was held to "get to the bottom of this frenzy," at which a person dressed up in alien costume with an oversized headpiece appeared, much to the laughter of the gathered press. Years later, however, during an appearance on History Channel's *UFO Hunters*, Symington apologized for the prank and admitted that he, too, had seen the giant floating triangle, hovering so close to the ground he could have hit it with a rock. But at the end of that interview, Symington also revealed that after the dust had settled from the event President Bill Clinton had given him a full presidential pardon for crimes Symington had been charged with when he was in office. A strange turn of events.

George W. Bush's vice president, Dick Cheney, was said to be among the most knowledgeable about UFOs. When he and Bush were on the campaign trail, Bush was asked about UFOs and in response jerked his finger at Cheney, telling the questioner, "If anyone knows, he does."

For his part, Cheney told questioners, "If I had been briefed about UFOs, that briefing would have been classified, which means I can't talk about it." Notice that Cheney did not deny having been briefed about UFOs or that UFOs were only a figment of the national imagination.

President Barack Obama consistently joked about UFOs and Roswell, especially when he sought to make his critics sound ludicrous. And for her part, Democratic candidate Hillary Clinton has promised, on more than one occasion, to reveal the truth about UFOs as long as that truth does not compromise national security. It should be noted that her campaign chairman, John Podesta, who served in Bill Clinton's administration, was a part of the UFO disclosure lawsuit against NASA to reveal their files concerning the Keckburg UFO crash.

# Who Are the Men in Black?

### The Men in Black.

After the blockbuster movie and Will Smith's infectious title song, it is almost impossible to say those words without seeing in our mind's eye any of the dozens of memorable scenes from the movie—with Rip Torn, Will Smith, and Tommy Lee Jones in their black suits and white shirts standing in the foreground.

"I make this look good."

Indeed.

But the history of the "Men in Black" extends all the way back to Kenneth Arnold's 1947 "saucer" sighting and reports of their mysterious appearances—and equally mysterious disappearances—have permeated the legend and lore of UFOlogy for the past five decades.

Here are the most commonly believed facts about the Men in Black.

- The Men in Black usually travel in groups of three and appear shortly after someone reports a UFO sighting or an abduction experience.

- MIB always wear black suits. Sometimes the suits are impeccably clean and pressed; other times the suits have been reported as being dirty and wrinkled.

- MIB usually interview a subject about his or her UFO experience and then caution them against talking about their encounter with anyone else.

- Sometimes the MIB speak with heavy accents; other times they use very formal language, or slang or gangster lingo.

- MIB often show an intense interest in everyday (terrestrial?) objects, such as pens or forks.

- Some of the people visited by MIB have reported being physically threatened by them, although there does not seem to be any evidence or reports of anyone being harmed by one of the MIB.

- One of the rumors about the MIB is that they are aliens and that their job is to monitor and stifle reports of UFOs and alien contact so that other aliens can continue their activities on Earth without interference.

- The earliest account of a visit from one of the men in black was reported in 1947 by Harold Dahl. Dahl claimed that he had witnessed a UFO discharging "metallic" substances into the ocean off the coast of Maury, Washington and that the following morning he was visited by a stranger clad in a dark suit who hinted that Dahl and his family would be harmed if he told anyone about what he had seen. This entire story was ultimately revealed to be an elaborate hoax, and yet conspiracy theorists embraced the story and charged a cover-up.

- In September 1953, Albert Bender, founder of the International Flying Saucer Bureau, claimed that he was visited by three men in dark suits after he confided a UFO theory to one of the people with whom he corresponded. Bender's story was the true beginning of the Men in Black legend and became known in UFOlogy as the "Bender Mystery."

- Many of the MIB drove black Cadillacs or very old automobiles in pristine condition, leading to the belief that the cars were "created" especially for the MIB to use to blend in with humans as they performed their mysterious duties.

- MIB often wear heavy makeup, lipstick, and sometimes have strange Asian-like, or what some witnesses have described as vaguely "European," facial features.

- MIB often identify themselves as government agents.

- Many witnesses use similar language to describe their feeling that the MIB they spoke with seemed to be uncomfortable with their human bodies, and that they acted as though they were wearing something that didn't quite correctly fit.

- The incredibly successful 1997 movie *Men in Black*, starring Will Smith and Tommy Lee Jones, put a different spin on the MIB story. In the movie, the Men in Black were not aliens. They were human agents whose duty it was to monitor alien activity on Earth and in nearby galactic space.

# Are Cattle Mutilations Connected to UFOs?

The following facts were compiled by the FBI during their investigations into strange incidences of cattle mutilations.

- UFOs are often reported prior to the discovery of mutilated cattle and other animals.
- Some of the UFOs reported have been lights, black disks, and strange, ultra-quiet black helicopters.
- Usually, only cattle and horses are found mutilated.
- Most often, the animals' sexual organs are taken, including the testicles, and rectum.
- There have been cases of mutilations where the animals' heart, liver, kidneys, udder, muscle fibers, womb, brain, eyes, tongue, nostrils, lips, and complete lower jaw have also been taken.
- In some cases, every bone in the animal's body has been broken. Some claim this is from being dropped from a

spacecraft from a high distance after the mutilation has been carried out.

- Near the site of mutilations, there are often circular imprints in the ground which register as highly radioactive. Some say these imprints are from spacecraft landing legs.

- The mutilated animals are often found to be marked on one side with a potassium and magnesium powder. This powder can only be seen under ultra-violet light.

- Some researchers into cattle mutilations believe that the removal of animal body parts is part of a long-term program of biological and genetic study of terrestrial life forms by visiting alien entities.

- Skeptics believe that all mutilations can be explained as either hoax, attacks by other animals, or secret government study of Transmissible Spongiform Encephalopathies among infected cattle.

# 25 Facts about the Face on Mars and Other Monuments on the Red Planet

*That the Face itself could also have an esoteric or theosophical meaning is certainly possible, but it will not redefine our entire Gnostic and dharmic heritage; it will simply translate terrestrial terms into interplanetary or transgalactic terms, perhaps replacing astral symbology with actual cosmic geography. But the Face (as artifact) at very least must redefine the exoteric history and "anthropology" of our planet.*

—Richard Grossinger in the "Publisher's Foreword" to Richard C. Hoagland's book *The Monuments of Mars: A City on the Edge of Forever*

The Face on Mars is one of the most fascinating subsets of the UFO phenomenon. Books have been written about it, documentaries have been produced about it, and lectures about the Face and the other Monuments on Mars always attract huge, passionately interested crowds.

This list looks at some of the facts surrounding the discovery of the image of the Face, the possibility that Mars may have once been inhabited, and the new 1998 NASA findings on the Martian terrain and its mysterious Monuments.

1. The first pictures of the mesa formation that would become known to the world as the Face on Mars were taken on Sunday, July 25, 1976, by the unmanned spacecraft Viking Orbiter 1 as it flew 1,100 miles above the surface of Mars.

2. The photo of the Face was NASA Frame 35A72, and it was taken during Viking's 35th orbit around Mars.

3. The Face is located in the northern hemisphere of Mars at approximately 41 degrees north latitude and 9.5 degrees west longitude.

4. The first person to notice the Face was Viking project scientist Tobias Owen, a member of NASA's imaging team.

5. Tobias Owen's first words after seeing the Face were, "Oh my God, look at this!"

6. Viking project scientist Gary Soffen was the NASA official who initially spoke to the media about the Face. He told them, "Isn't it peculiar what tricks of lighting and shadow can do. When we took a picture a few hours later it all went away. It was just a trick, just the way the light fell on it."

7. Experts now believe that the reason the Face "went away" (in the words of NASA spokesman Gary Soffen) a few hours after the initial photos were taken was not because the Face had been an illusion, but because the Martian night had fallen.
8. On July 31, 1976, NASA released the first photos of the Face and issued the following press release:

*This picture is one of many taken in the northern latitudes of Mars by the Viking 1 Orbiter in search of a landing site for Viking 2.*

*The picture shows eroded mesa-like landforms. The huge rock formation in the center, which resembles a human head, is formed by shadows giving the illusion of eyes, nose and mouth. The feature is 1.5 kilometers (one mile) across, with the sun angle at approximately 20 degrees. The speckled appearance of the image is due to bit errors, emphasized by enlargement of the photo. The picture was taken on July 25 from a range of 1873 kilometers (1162 miles). Viking 2 will arrive in Mars orbit next Saturday (August, 7) with a landing scheduled for early September.*

9. The Face on Mars has been said by some to resemble the Sphinx from ancient Egypt, leading inevitably to questions about the possible extraterrestrial origin of the Egyptian monuments and the connection between the monuments on Mars and Earth history.

10. Some believe that there was a horrible cataclysm on Mars that destroyed all civilization on the red planet and that the Face and the other Monuments were constructed as a warning to Earthlings that we may be in line for the same type of catastrophic destruction.

11. After discovery of the Face on Mars, the Cydonia region of Mars was declared unfit to be used as a landing area for Viking 2. NASA said it was too hazardous, although the landing region ultimately chosen, Utopia, was deemed by NASA to be as dangerous for Viking 2 to land in as Cydonia, if not even more dangerous.

12. Several miles southwest of the Face is a set of pyramidal structures that some believe are the abandoned components of an engineered Martian city.

13. People who believe the Face on Mars is a sure sign that Mars was once inhabited point to several similarities between Mars and Earth, including:

    - Mars' axis is tilted at an angle of 24.935° in relation to the plane of its orbit around the Sun; the Earth's axis is tilted at an angle of 23.5° in relation to the plane of its orbit around the Sun.
    - Mars makes a complete rotation around its axis in 24 hours, 39 minutes, 36 seconds; the Earth makes a complete rotation in 23 hours, 56 minutes, 5 seconds.
    - Mars and the Earth both manifest a cyclical wobble of their axis known as precession.
    - Mars and the Earth both have the same non-spherical shape: Their poles are flattened, and their equators bulge.
    - Mars has four seasons; Earth has four seasons.
    - Mars has icy polar caps; Earth has icy polar caps.
    - Mars has deserts; Earth has deserts.
    - Mars has dust-storms; Earth has dust-storms.
    - Scientists have calculated that Mars' surface temperature was at one point during its existence about the same as that of the Earth's now.

14. In 1985, computer imaging analysis of photos of the Face by Mark Carlotto identified decorative crossing lines above the eyes that suggest a crown; teeth in the mouth of the object; and a very Egyptian-like striped headdress, similar to the ones worn by Egyptian pharaohs on Earth.

15. Richard Hoagland, author of *The Monuments of Mars*, asked the following rhetorical question about the possible existence on Mars of engineered Monuments, including the Face: "What better way to call attention to a specific place on Mars as a site for further exploration than by using a humanoid image?"

16. According to astronomical calculations performed by Dr. Colin Pillinger of the UK Planetary Sciences Research Institute, up to 100 tons of Martian material—in the form of meteorites and other Martian rocks—land on Earth each year.

17. Mars is now a dead planet. Its average temperature is -23° Celsius, and water cannot survive there. On Mars, there is only ice. And yet there is geologic evidence (found in recovered Martian meteorites and on the Martian terrain) of enormous floods as recently as 600,000 years ago. And if there were floods, then Mars would have supported the existence of water, which, by extension, would have supported the existence of life. One recovered Martian meteorite actually contained a droplet of real water.

18. Other geological analysis of the surface of Mars points to evidence of ancient shorelines, all of which are quite close to the Cydonia region and the Face on Mars.

19. Within ten miles of the Face is a five-sided pyramid now known as the D&M Pyramid, named after its discoverers Vincent DiPietro and Gregory Molenaar.

20. Other structures on Mars that appear to be engineered include monuments christened The Fort, The Cliff, and The City.

21. Richard Hoagland used computers to come up with an astonishing fact. He discovered that a viewer standing at the center of the configuration of monuments known as The City would have seen the sun rising out of the mouth of the Face at dawn on the Martian summer Solstice 330,000 years ago.

22. NASA continues to insist that all of the so-called "Monuments on Mars" are completely natural geologic formations. In response to this assertion, California Institute of Technology Professor of Geology and Planetary Science Arden Albee said, "[A]s of yet, there is no natural geological explanation for the Cydonia Structures."

23. In April 1998, NASA instructed its latest orbiter, Mars Global Surveyor, to make three passes over the Cydonia region. The first pass on April 5 captured the Face with astonishing accuracy—although at first, the image appeared to be nothing more than a flat ridged area of the Martian terrain. After computer analysis and enhancement, it became obvious that the photo had been taken through clouds and was under-processed. Mark Carlotto's computer imaging work with the photos later revealed even more detail of the Face, including nostrils.

24. Former NASA consultant and acclaimed astronomer Dr. Tom Van Flanders studied the new photos and issued the following statement: "The humanoid facial features that first drew attention to this area are confirmed by this photo, despite poor lighting and poor viewing angle. Using the ability to change mental perspectives, one can see the object clearly, without imagining details, as an excellent rendition

of a sculpted face. In my considered opinion, there is no longer room for reasonable doubt of the artificial origin of the face, and I've never concluded 'no room for reasonable doubt' about anything before in my thirty-five-year scientific career."

**25.** Of the debatable quality of the second batch of NASA'S Martian Cydonia photos, Richard Hoagland smelled a cover-up and stated the following on April 6, 1998, on the Art Bell radio program: "There is a picture tonight which is, yes, geographically and geometrically of the Face on Mars. Now, it's not anywhere near the kind of picture we should have gotten. It is light years below the level this camera and this technology is capable of giving us, so they're giving us the business, but it is a picture geometrically targeted correctly. So the first part of what I would want has happened. Now we know they can target. Now there's no reason they can't target the important stuff, which is the geometry in the city, the Pyramids, the material that is numerically testable, and that's what we should be demanding, and oh, by the way . . . take the lens cap off the camera. TELL THE MEDIA TO FACE THE TRUTH!"

# 23 Facts about the Crop Circles Phenomenon

*How would an extra-terrestrial civilisation effectively communicate with us without being instantly silenced by scientists and national armies?*

—Erich von Däniken, *Chariots of the Gods*

Crop circles are one of the more unusual aspects of UFOlogy. They appear without warning and are beautiful and strange. Here is a look at some of the facts surrounding these odd agrarian occurrences.

1. Crop circles are beautiful, complex, geometric patterns that mysteriously appear in crop fields overnight and which can only be seen in their entirety from the air.

2. There are written reports—but no photographic evidence—of crop circles dating back to the 1930s and 1940s.

3. "Crop" circles have appeared in wheat, winter barley, rye, oilseed rape, oats, rice, vegetables, groups of trees, and in the snow.

4. According to crop circle authority Michael Hesemann, explanations for crop circles include UFO landings, earth

spirits, stationary whirlwinds, morphogenetic fields, the collective unconscious of humanity, the telekinetic powers of crop circle researchers, pranksters, geomagnetic fields, and copulating hedgehogs.

5. Crop circles range in size from three feet across to over 500 feet across.

6. The average size of a crop circle is approximately 150 feet across.

7. One popular theory regarding the appearance of crop circles is that they are a means of preparing us for extraterrestrial visitation. If mankind gets used to messages from aliens presented in an aesthetically pleasing, environmentally friendly manner, the thinking goes, then when the ETs land, we might be more willing to welcome them instead of annihilating them with surface-to-air missiles as soon as they enter our airspace.

8. Crop circles appear mostly between late May and September in the Northern Hemisphere.

9. It is not uncommon for fields with circles to be revisited at a later time, with new circles appearing next to the originals, or the older circles expanded with new patterns.

10. Tape recorders often pick up a 5 kilohertz buzzing sound within crop circles.

11. Crop circles became increasingly complex in the fourteen years between 1978 and 1992.

12. UFO sightings and reports often precede the appearance of a crop circle.

13. When circles appear in corn fields, the stems of the corn plants are always bent, never broken.

14. Crop circles often appear in the vicinity of ancient burial mounds or near important sites of prehistoric archaeology.

15. A crop pictogram found at Milk Hill above Alton Barnes, Wiltshire in England in August 1991 was determined to be a form of an ancient Hebrew language that translated to "The Creator, wise and kind."

16. Some crop circle pictograms in England point with amazing accuracy to Stonehenge and other ancient sites such as Silbury Hill.

17. For years, the CIA collected information about crop circles.

18. So far, crop circles have appeared in Ireland, Sweden, the Netherlands, England, Germany, Canada, Belgium, France, Spain, Switzerland, Romania, Hungary, Bulgaria, the Czech Republic, the Soviet Union, Australia, New Zealand, Japan, Afghanistan, Turkey, Egypt, Brazil, Puerto Rico, Mexico, and the United States.

19. On the night of Sunday, June 16, 1991, the night the "mother of all pictograms" appeared at Bradbury Castle in England, pulsing UFOs were seen overhead by three people, Brian Grist, Gary Hardwick, and Gary's girlfriend, Alison.

Of their close encounter, Brian later commented, "It remind[ed] me most of scenes in the film *Close Encounters of the Third Kind*. I simply couldn't believe what was happening there." The next morning, the enormous pictogram was first discovered.

20. According to British psychic Isabelle Kingston (who has correctly predicted the sites of new crop circle appearances), an "intelligence outside this planet" known as the Watchers have guided humankind's fate since ancient times and are responsible for the formation of the crop circles. The Watchers told Kingston that the circles were a means of preparing humankind for a "new age" of enlightenment.

21. According to "Face on Mars" authority Richard Hoagland and others, the mathematical values of the "Mother of All Pictograms" at Bradbury Castle correlate precisely to the distances, values, and coordinates of the "Martian city" in the Cydonia region of Mars, the location of the Face on Mars.

22. The formation of crop circles has also been associated with incidences of cattle mutilations.

23. True believers in an extraterrestrial genesis for crop circles admit that up to 30 percent of existing circles may be fakes—but point to the remaining 70 percent of the circles worldwide as unassailable proof of a *non-terrestrial* cause for the manifestations.

# The Hollow Earth Theory & UFOs

*It is [my] purpose to present scientific evidence to prove that the Earth, rather than being a solid sphere with a fiery center of molten metal, as generally supposed, is really hollow, with openings at its poles. Also, in its hollow interior exists an advanced civilization which is the creator of flying saucers.*

—Dr. Raymond Bernard, *The Hollow Earth: The Greatest Geographical Discovery in History*

The Hollow Earth Theory is one of the more fanciful theories of UFOlogy.

As described in the Foreword cited above, Hollow Earth theory states that the Earth is really a hollow sphere and inside this massive globe are rivers, mountains, and forests. Most significantly, intelligent civilizations—"super races"—reside inside the globe, and are the source of many of the UFO sightings seen in the skies above our planet.

As with fringe UFO theories like Bigfoot and others, there are also Hollow Earth "experts," and one of the most notable is the above-quoted Dr. Raymond Bernard, author of a most momentously-titled book.

Science has proven that the Earth's core is molten and that life cannot survive *inside* the planet.

And yet, the Hollow Earthers cite all kinds of evidence denying these findings and wholeheartedly believe that the world governments know the truth and that the "scientific findings" about the Earth's molten core are nothing but government disinformation.

Here are two excerpts—the thirteen principles—that spell out the specifics of this bizarre theory. This list is from Dr. Bernard's 1969 book and is titled, "What This Book Seeks To Prove."

1. That the Earth is hollow and not a solid sphere as commonly supposed, and that its hollow interior communicates with the surface by two polar openings.

2. That the observations and discoveries of Rear Admiral Richard E. Byrd of the United States Navy, who was the first to enter into the polar openings, which he did for a total distance of 4,000 miles in the Arctic and Antarctic, confirm the correctness of our revolutionary theory of the Earth's structure, as do the observations of other Arctic explorers.

3. That, according to our geographical theory of the Earth being concave, rather than convex at the Poles, where it opens into its hollow interior, the North and South Poles have never been reached because they do not exist.

4. That the exploration of the unknown New World that exists in the interior of the Earth is much more important than the exploration of outer space; and the aerial expeditions of Admiral Byrd show how much exploration may be conducted.

5. That the nation whose explorers first reach this New World in the hollow interior of the Earth, which has a land mass greater than that of the Earth's surface, which may be done by retracing Admiral Byrd's flights beyond the hypothetical North and South Poles, into the Arctic and Antarctic polar openings, will become the greatest nation in the world.

6. That there is no reason why the hollow interior of the Earth, which has a warmer climate than the surface, should not be the home of plant, animal, and human life. If so, it is very possible that the mysterious flying saucers come from an advanced civilization in the hollow interior of the Earth.

7. That, in event of a nuclear world war, the hollow interior of the Earth will permit the continuance of human life after radioactive fallout exterminates all life in the Earth's surface; and will provide an ideal refuge for the evacuation of survivors of the catastrophe, so that the human race may not be completely destroyed, but may continue.

To paraphrase Vincent Vega from *Pulp Fiction*, "Those are some bold statements!" After stating his intentions with the above seven points, Dr. Bernard then proceeds to provide proof of his theories, utilizing scientific fact and theory, ancient writings, and NASA photographs. Dr. Bernard concludes his book by summing up his theory with the following six points:

8. There is no North or South Pole —where they are supposed to exist there are actually wide openings to the hollow interior of the Earth.

9. Flying saucers come from the hollow interior of the Earth through these polar openings.

10. The hollow interior of the Earth, warmed by its central sun (the source of Aurora Borealis) has an ideal subtropical climate of about 76 degrees in temperature, neither too hot nor too cold.

11. Arctic explorers found the temperature to rise as they traveled far north, they found more open seas, animals traveling north in winter seeking food and warmth when they should have gone south. They found the compass needle to assume a vertical position instead of a horizontal one and to become extremely eccentric. They saw tropical birds and more animal life the further north they went; they saw butterflies, mosquitoes and other insects in the extreme north when they were not found until one is as far south as Alaska and Canada. Additionally, they found the snow discolored by colored pollen and black dust, which became worse the further north they went. The only explanation is that this dust came from active volcanoes in the polar opening.

12. There is a large population inhabiting the inner concave surface of the Earth's crust, composing a civilization far in advance of our own in its scientific achievements, which probably descended from the sunken continents of Lemuria and Atlantis. Flying saucers are only one of their many achievements. It would be to our advantage to contact these Elder Brothers of the human race, learn from them, and receive their advice and aid.

13. The existence of a polar opening and land beyond the Poles is probably known to the US Navy in whose employ Admiral Byrd made his two historic flights, and which is probably a top international secret.

Excerpts from *The Hollow Earth: The Greatest Geographical Discovery in History* (Citadel Press) by Raymond Bernard. ©1969 University Books Inc. Used by permission. All rights reserved.

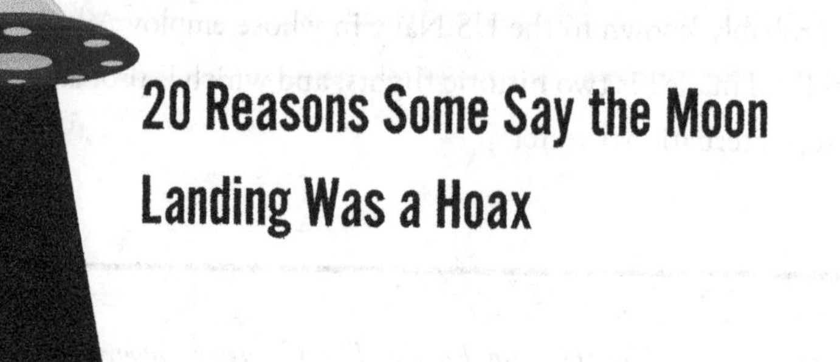

# 20 Reasons Some Say the Moon Landing Was a Hoax

*As for the Yankees, they had no other ambition than to take possession of this new continent of the sky, and to plant upon the summit of its highest elevation the star-spangled banner of the United States of America.*

—Jules Verne, *From the Earth to the Moon*

Some people believe that there are alien bases on the Moon.

Some people believe that the Moon is the source of UFOs seen in the skies above our planet.

And some people believe that the Moon has never been walked on by human beings and that the whole *Apollo* landing mission was faked.

Here is a look at what *these* people believe.

1. NASA's July 20, 1969, *Apollo* landing on the Moon in the Sea of Tranquility never happened.

2. None of the five subsequent Moon landings took place either.

3. All the Moon landings were faked (with help from the Walt Disney Studios) on a top-secret movie soundstage at the

Atomic Energy Commissions site in Nevada, 90 miles north of Las Vegas.

4. Elements of the $30 billion hoax included faked photographs, fake moon rocks, well-rehearsed astronauts, and the collusion of Walter Cronkite.

5. Supporters of the hoax theory point to the fact that there were absolutely no stars visible in the lunar sky in any photographs taken on the moon. Without an atmosphere, the celestial panoply would have been breathtaking, and the reason there are no stars is that NASA's set decorators knew they'd never even be able to come close to what the sky would look like, so they left it pitch black.

6. If the surface of the moon was powdery enough for the astronauts to leave footprints, why didn't the lunar lander dig a crater when it landed?

7. Why is there no moon dust on the lunar lander's legs?

8. Why were astronauts quarantined after "returning" to Earth if the Moon was confirmed to be absolutely sterile?

9. Many astronauts who went to the Moon ended up with high-paying positions with major corporations.

10. The reason NASA faked the Moon landing was that everyone at the space agency knew that they did not have the technology or expertise to land a man on the Moon by the end of the sixties. President Kennedy had promised the landing, and they wanted to save the United

States from being embarrassed and denigrated around the world.

11. According to hoax theorists, NASA worked with the Las Vegas "Cosa Nostra" to perpetrate the hoax.

12. Director Stanley Kubrick helped out with special effects he had developed for his movie *2001: A Space Odyssey*.

13. The Saturn V rocket that lifted off in Florida for the first Moon mission was actually empty and, as soon as it was out of sight of the public and their telescopes, the empty rocket was sent plummeting into the South Polar Sea.

14. According to hoax theorists, lunar exploration missions stopped because NASA ran out of pre-filmed episodes and everyone in the know at the space agency felt that they were dangerously close to getting caught.

15. While the lunar spacecraft was on its way to the moon, the *Apollo* astronauts were allegedly partying hearty in Las Vegas.

16. The moon rocks were made in a kiln at NASA.

17. When it was time for the astronauts to "return to Earth," they were jetted to an air base on the Tauramoto Archipelago south of the Hawaiian Islands. There they were sealed into a fake capsule, flown out to the drop point, and dropped into the ocean from a C5-A transport plane.

18. Some hoax theorists believe that the entire Moon landing "performance" was a metaphorical illustration of the ancient

Illuminati religion and that all the astronauts are actually Freemasons who subscribe to the occult practices and beliefs of the religion.

**19.** Supposedly, there is a photograph in the Masonic House of the Temple in Washington, D. C. of Neil Armstrong standing in his spacesuit on the "moon's" surface holding a Masonic Apron in front of his groin.

**20.** The political/national security strategy for the faked moon landing was to goad the USSR into outspending us on the space race to get to the moon so that it would bankrupt itself in the process. President Nixon believed that the Soviets were in no position to get into a spending war with the United States. Thus, his plan was to force them into matching our defense spending and space exploration strategy dollar for dollar. By faking the moon landing, Nixon hoped to trick the Soviets into buying something they couldn't afford: the moon.

# PART IX

# POPULAR CULTURE AND THE UFO PHENOMENON

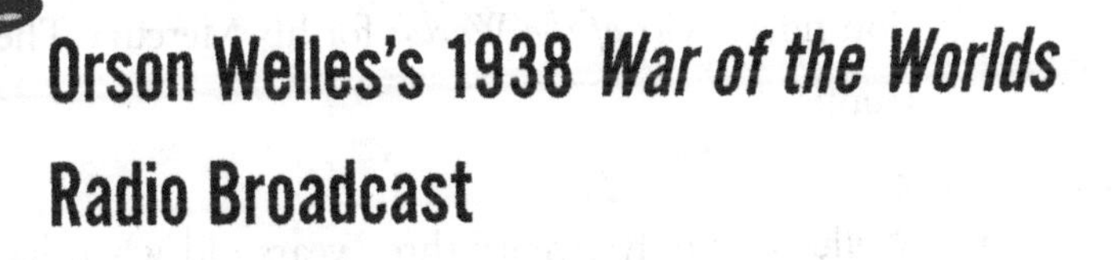

# Orson Welles's 1938 *War of the Worlds* Radio Broadcast

Many knowledgeable UFOlogists point to the panic that ensued because of Orson Welles's *War of the Worlds* broadcast in 1938 as the beginning of the US's (and other countries') deliberate attempt to withhold accurate UFO information from the general public.

The governments around the world looked at how the "masses" reacted when they believed that the world was being invaded by Martians and, from that point on, it is asserted, a deliberate policy of disinformation and secrecy was adopted by governments everywhere.

The average person cannot handle the truth, the purported bureaucratic thinking went, and, thus, the facts about UFOs, Roswell, Area 51, MJ-12, Monuments on Mars, abductions, cattle mutilations, crop circles, and all other manner of bizarre phenomena, while undoubtedly known at the highest levels of government, will never be made available to the general public.

There is a fatal flaw in this thinking, however.

There is a significant difference between scaring the bejesus out of a population with a phony news broadcast and making available facts and data about a scientific reality. But, as is often typical for government agencies and bureaucrats, official policy was adopted shortly after the 1938 panic and has never been changed.

This feature looks at the details surrounding this seminal moment in the history of UFOs and illustrates the reason why important information would be withheld from people around the world, from 1938 until today.

- On Sunday, October 30, 1938, the day before Halloween, Orson Welles dramatized H. G. Wells's classic science fiction novel *War of the Worlds* for his Mercury Theater radio troupe.

- Welles was only twenty-three years old when he produced the radio program that terrified millions of people and caused a national panic.

- A Princeton University study later determined that a total of 6 million people heard Welles's broadcast. Of those listeners, 1.7 million believed it was a real news bulletin, and 1.2 million actually took action because of it.

- Welles co-wrote the script for the broadcast with Howard Koch, who would later go on to win an Oscar for writing *Casablanca*.

- The *War of the Worlds* broadcast began as a normal radio program featuring "Ramon Raquelo and His Orchestra from the Hotel Park Plaza."

- After a few minutes of music, "Newsman Carl Phillips" broke into the broadcast, telling listeners, "We interrupt our program of dance music to bring you a special bulletin from the Intercontinental Radio News," continuing with the alarming news that a "large, flaming object" had fallen from the sky onto a farm in Grover's Mill, New Jersey.

- Welles deliberately scheduled the "news bulletin" to occur at twelve minutes into the broadcast, knowing that millions of listeners would be tuning their radio dials away from the popular *Chase and Sanborn Hour* (after Edgar Bergen and Charlie McCarthy had done their shtick.) Meaning many would come across his program well after the disclaimer that the show was only a dramatization.

- According to Newsman Carl Phillips, the Martians that landed in Grover's Mill were repulsive to the max: "Good heavens, something's wriggling out of the shadows like a gray snake. I can see the thing's body. I can hardly force myself to keep looking at it. The eyes are black and gleam like a serpent. The mouth is kind of V-shaped with saliva dripping from its rimless lips."

- For forty-five minutes, the invading Martians burned down Grover's Mill and also killed Carl Phillips as they headed for New York City, destroying everything in their path.

- The New York City police department received two thousand hysterical emergency calls from people during the first fifteen minutes of the broadcast.

- The *New York Post*'s headline the following day read, "Radio 'War' Panic Brings Inquiry; US To Scan Broadcast Script. Mars 'Attack' Brings Wave of Hysteria." On their editorial page, the *Post* called Welles's actions "unscrupulous and irresponsible."

- Sociocultural scholars suggest that Welles's broadcast ended a period of American innocence and trust and ushered in a time of distrust and paranoia. Welles seemed to confirm that that was his intention all along: "The radio was believed in America," he said shortly before he died. "That was a voice from heaven, and I wanted to destroy that as dramatically as possible. And that was the purpose of this huge practical joke, which was much bigger than we expected it to be."

- The national panic aside, Welles's program resulted in him signing a film deal with RKO Pictures. Two years later, he picked his first project: *Citizen Kane.*

All of this having been said, there is another truth that belies the story of the panic over the 1938 Mercury Theater of the Air Broadcast.

- The entire program was paid for by the Rockefeller Foundation as part of a larger study in analytics managed by Frank Stanton at Princeton University. Thus, the reason that the Martians landed in Grovers Mill, which is right outside of Princeton, New Jersey.

- Stanton's study funded by the Rockefellers was designed to assess how mass populations could be manipulated via mass media, a study that has had profound influences on popular culture over the past ninety years.

- There actually was no "panic in the streets," as Frank Stanton himself revealed to British newspapers. In fact, Stanton said that he was so disappointed in the public reaction, he falsified articles describing panic in the streets to bolster his argument that mass populations could be manipulated.

- Part of the rationale for the Rockefeller/Stanton study was a reaction to the two major cultural/political revolutions that overthrew governments earlier in the century, the

Communist Revolution in Russia and rise of National Socialism in Italy and Germany that brought the world to the brink of war in 1938. In both instances, inherently undemocratic governments rose to power by using the media to manipulate mass populations. The *War of the Worlds* broadcast was an attempt to study the power of mass media to spread fear throughout a population.

# UFO Articles from *Fate* Magazine from the 1960s

*Fate* magazine has been around since 1948, the year after Kenneth Arnold saw the flying objects he described as "saucers," and, in a time when magazines start up, publish a few issues, and go belly up, *Fate* is coming up on seven decades of continuous publication.

*Fate*'s first issue (Spring 1948) cost twenty-five cents and included the articles, "The Truth About Flying Saucers," "Invisible Beings Walk the Earth," "Mark Twain and Haley's Comet," and "Twenty Million Maniacs."

One of *Fate*'s recent issues cost $3.50 and included the articles, "To Death and Back: Children Share Their Glimpses of the Afterlife," "Brazil's UFO Cult," "My Search for Past Lives," "Spirit Guide: A Primer of Otherworldly Beings," "Leslie Lang: Police Psychic," and "Religious Roots of Reincarnation."

Isn't is nice to know some things never change?

*Fate* is available on newsstands or by calling 1-800-THE-MOON to subscribe.

## FEBRUARY 1962

Cover Headline:

SPECIAL UFO SECTION!

24 pictures . . . 20 pages . . . study of

281 photos and 411 flying saucers

- "My First UFO" by Frank Edwards
- "Camera's Eye Analysis of Flying Saucers" by George D. Fawcett

## JUNE 1965

Cover Headline:

UFOs – MEASURING, WEIGHING, ANALYZING

- "Scientific Measurement of UFOs" by Professor Charles A. Maney

## JANUARY 1967

Cover Headline:

UNITED NATIONS HEARS THE CASE:
THE BARNEY HILL MYSTERY:
KIDNAPPED BY A UFO

- "Kidnapped By a UFO" by Allen Spraggett
- "What Can We Expect of 'Civilization' in Outer Space?" by James Crenshaw
- "Open Letter" by Dr. J. Allen Hynek

## APRIL 1967

Cover Headline:

FLYING SAUCERS ARE REAL

- "Flying Saucers Are Real" by Jacques and Janine Vallee

## MAY 1967

Cover Headline:

WAS MARINER IV SABOTAGED BY A UFO?

- What happened to the instruments on the Spacecraft while it was "delayed" for eight minutes behind the red planet Mars?
- "Did a UFO Sabotage Mariner IV?" by Randall C. Hecker

## AUGUST 1967

Cover Headline:

SPECIAL: What kind of saucer have you seen? Let's play THE UFO NUMBERS GAME

- "UFO Numbers Game" by John Weigle

## FEBRUARY 1969

Cover Headline:

UFOs OVER ITHACA, N.Y.

- "UFOs Over Ithaca" by T. M. Wright

# Stephen King Stories about UFOs or Aliens

Stephen King is known around the world as the most famous and most published horror writer of the twentieth century.

And yet, King is much more than "just" a horror writer—as this list of UFO-themed Stephen King stories and novels illustrates.

**NOTE:** These story and novel synopses (which are in chronological order) are adapted from my 1998 book, *The Lost Work of Stephen King.*

### "I Am the Doorway" (*Night Shift*)

A deep-space astronaut named Arthur returns from a mission to Venus "infected" with alien creatures who burrow up through the skin of his fingers and grow eyes with which they view our world.

The crippled and wheelchair-bound Arthur (he was seriously injured during re-entry) is the "doorway" for these creatures who can control his body at will and make him kill at their command.

"I Am the Doorway" was the story chosen as the source for the cover illustration for the Signet paperback edition of *Night Shift.* The (uncredited) cover drawing shows a right hand half wrapped (unwrapped, actually) in gauze; the top half of the hand revealing eight beautiful blue eyes scattered across the fingers and palm.

"I Am the Doorway" has all the trappings of science fiction—spaceships, travel to other planets, malevolent aliens, etc., and yet King deftly uses these genre elements to write an unabashed *horror* story, brilliantly blurring the barrier between the two genres and creating the memorable hybrid genre of sci-fi/horror.

"I Am the Doorway" was written in 1971, and King postulates Earthlings landing on Mars in 1979. As you know, man did finally land a controllable vehicle on Mars in 1997—the exploration craft *Sojourner*—but as of yet, a manned mission to the Red Planet has not been possible.

King has Arthur fearing that the early manifestations of his "infestation" is actually leprosy, utilizing the theme of the horrors of physical "decay"—a leitmotif that he will return to again and again in his work. King will later make very effective use, for instance, of the metaphor of cancer being an alien creature eating a person up from the inside out. In "I Am the Doorway," he has the enemy alien creatures *living* inside Arthur and using him as a host for their murderous deeds.

### "The Lonesome Death of Jordy Verrill" (*Creepshow*)

The story is classic sci-fi-horror: A meteor from outer space lands on Jordy Verrill's New Hampshire farm one Fourth of July evening. The meteor contains a sentient, vegetative alien life form that begins to propagate by growing weeds on anything it touches. It can feed on the earth, thrives when given water, and within hours after it has landed, it is already blooming. Witless Jordy touches the meteor and begins transforming into a walking plant. When he is literally nothing but a green monster, he puts himself out of his misery by blowing his own head off with a shotgun. The story ends with a weather forecast calling for rain. The last lines of the short story consist of the thoughts of the alien weeds:

> *Jordy-food.*
> *A fine planet, a wet planet. A ripe planet.*
> *Cleaves Mills-food.*
> *The weeds began to grow toward town.*

### "Paranoid: A Chant" (*Skeleton Crew*)

This precisely 100-line poem takes us inside the mind of a man who has seen UFOs and is utterly mad and completely paranoid.

The poem's narrator sees people watching him and knows that "Men have discussed me in back rooms." His mother has been investigated, the FBI is monitoring him, and the old woman who lives upstairs from him is radiating him with a suction cup device attached to her floor. His mail contains letter-bombs, waitresses are poisoning his food, and "a dark man with no face" has surfaced in his toilet to listen to his phone calls.

King effectively communicates madness with his superb use of language and disjointed free-form lines; taking us inside the mind of a man who has created a bizarre, fully-realized interior reality that must, for the reader, co-exist with the mundane details of the narrator's *exterior* reality. This is a literary high-wire act and yet King pulls it off masterfully.

King uses some very powerful surrealistic images in this poem, especially that of the black crows with black umbrellas and silver dollar eyes standing at the bus stop looking at their watches.

### "Beachworld" (*Skeleton Crew*)

This is a fun story in which a living sand planet devours invading astronauts. King tells us that the planet itself is a sentient—and hungry—being, and it decides to eat two hapless US astronauts.

### *It*

*It* is King's magnum opus. This massive novel is a multiple timeline, multiple character epic that tells the history of the Maine town of Derry—and the timeless and evil thing that lives in its sewers. We and the main characters, known as The Losers, learn that the thing living under Derry is not from around here—in fact, it's from beyond the stars. For all its horrors and terrors, *It* must, in the end, be ranked as a great science fiction novel—albeit one with countless elements of nightmarish horror.

### *The Tommyknockers*

In this expansive science fiction novel, writer Bobbi Anderson finds an ancient alien spacecraft buried in the woods near her home in Maine and begins digging it up. Bobbi thinks the aliens are dead, but they have only been in deep stasis and, as she awakens them, they begin executing their horrible plan for domination.

### *The Langoliers*

In this novella, a red-eye flight from L.A. to Boston goes terribly wrong when the passengers realize that the US they know is no longer below them—and that there is something waiting for them in the dark. This "something" is devouring aliens who are capable of actually eating the very fabric of time and reality.

### "The Ten O'Clock People" (*Nightmares & Dreamscapes*)

The truly strange premise for this sci-fi/horror story is that horrible bat-like alien creatures with oozing tumors on their faces and a taste for humans (especially eyeballs it seems) have infiltrated the highest levels of American business and culture (the Vice President is one) with a plan to take over the world—ultimately using the human race as their own personal All-You-Can-Eat buffet. The only ones that can see these bat creeps in their true form are the "Ten O'Clock People."

And why is it that these certain folks—including the hero of this tale, Brandon Pearson—are gifted with the ability to see the bats' malformed, tumor-ridden, grayish-brown, hairy visages?

Because they are all Ten O'Clock People: smokers who are in that anxious, unpleasant area between smoking like a chimney and quitting completely. The Ten O'Clock People have restricted their smoking to between five and ten cigarettes a day, and that seems to be the intake range within which their brains develop the ability to see the bats.

King got the idea for "The Ten O'Clock People" while wandering around Boston one morning and noticing the groups of smokers gathered in front of their buildings having their "ten o'clock" nicotine fix.

### "The House on Maple Street" (*Nightmares & Dreamscapes*)

In this terrific short story, a wicked stepfather is sent into what King describes as the "Great Beyond" when his house literally *takes off into space* like a spaceship.

# 129 Books about UFOs, Extraterrestrials, and Other Examples of "High Strangeness"

The following is a UFO library that will serve you well if you are interested in learning more about one of the most puzzling and mysterious phenomena of the human experience.

1. *Abducted! The Story of the Intruders Continues* by Debbie Jordan and Kathy Mitchell. (Carrol & Graf, 1994).
2. *Abduction: Human Encounters with Aliens* by John E. Mack. (Scribner's, 1994).
3. *Above Top Secret: The Worldwide UFO Cover-Up* by Timothy Good. (Morrow, 1988).
4. *Alien Agenda: Investigating the Extraterrestrial Presence Among Us* by Jim Marrs. (HarperCollins, 1997).
5. *Alien Contact: The First Fifty Years* by Jenny Randles. (Sterling, 1997).
6. *Alien Contact: Top-Secret UFO Files Revealed* by Timothy Good. (Quill, 1994).
7. *Alien Contacts and Abductions: The Real Story from the Other Side* by Jenny Randles. (Sterling, 1994).
8. *Alien Discussions: Proceedings of the Abduction Study Conference* edited by Andrea Pritchard, David E. Pritchard, John E. Mack, Pam Kasey, and Claudia Knapp. (North Cambridge Press, 1994).

9. *Alien Encounters (Mysteries of the Unknown Series).* (Time Life, 1992).
10. *Alien Encounters* by Chuck Missler and Mark Eastman. (Harvest House, 1997).
11. *Alien Harvest: Further Evidence Linking Animal Mutilations and Human Abductions to Alien Life Forms* by Linda Moulton Howe. (Self published, 1993).
12. *Alien Identities: Ancient Insights into Modern UFO Phenomena* by Richard L. Thompson. (Govardhan Hill, 1995).
13. *Alien Impact* by Michael Craft. (St. Martin's, 1996).
14. *Alien Update* by Timothy Good, editor. (Avon, 1993).
15. *The Aliens and the Scalpel: Scientific Proof of Extraterrestrial Implants in Humans* by Dr. Roger K. Leir. (Granite 1998, 19xx).
16. *The Andreasson Affair: The Continuing Investigation of a Woman's Abduction by Alien Beings* by Raymond E. Fowler. (Wild Flower Press, 1994).
17. *The Andreasson Legacy* by Raymond E. Fowler. (Marlowe & Co., 1997).
18. *Angels and Aliens: UFOs and the Mythic Imagination* by Keith Thompson. (Addison-Wesley, 1991).
19. *Area 51: Nightmare in Dreamland* by Sean David Morton. (M. Evans & Co., 1997).
20. *Area 51: The Dreamland Chronicles* by David Darlington. (Henry Holt, 1997).
21. *Beyond Roswell: The Alien Autopsy Film, Area 51, and the US Government Cover-up of UFOs* by Michael Hesemann and Philip Mantle. (Marlowe & Co., 1997).
22. *The Bible and Flying Saucers* by Barry Downing. (Marlowe & Co., 1997).
23. *The Biological Universe: The Twentieth-Century Extraterrestrial Life Debate and the Limits of Science* by Steven J. Dick. (Cambridge University Press, 1996).

24. *Breakthrough: The Next Step* by Whitley Streiber. (Harper-Collins, 1995).
25. *Bringing UFOs Down to Earth* by Philip J. Klass. (Prometheus, 1997).
26. *Chariots of the Gods?: Unsolved Mysteries of the Past* by Erich Von Däniken. (Bantam, 1968).
27. *Charles Fort: Prophet of the Unexplained* by Damon Knight. (Doubleday, 1970).
28. *Close Encounters of the Fourth Kind: Alien Abduction, UFOs, and the Conference at M. I. T.* by C. D. Bryan. (Knopf, 1995).
29. *Close Extraterrestrial Encounters: Positive Experiences with Mysterious Visitors* by Richard J. Boylan and Lee K. Boylan. (Wild Flower Press, 1994).
30. *Communion: A True Story* by Whitley Streiber. (Morrow, 1987).
31. *The Complete Book of UFOs: An Investigation into Alien Contacts & Encounters* by Jenny Randles and Peter Hough. (Sterling, 1996).
32. *Confrontations: A Scientist's Search for Alien Contact* by Jacques Vallee. (Ballantine, 1990).
33. *Contact of the 5th Kind: The Silent Invasion Has Begun: What the Government Has Covered Up* by Philip J. Imbrogno. (Llewellyn, 1997).
34. *The Cosmic Connection: an Extraterrestrial Perspective* by Carl Sagan. (Dell, 1973).
35. *The Cosmic Connection: Worldwide Crop Formations and ET Contacts* by Michael Hesemann. (Gateway Books, 1996).
36. *Cosmic Test Tube: Complete Book of Extraterrestrial Contactories, and Evidence* by Randall Fitzgerald. (Moon Lake Media, 1997).
37. *Cosmic Voyage: A Scientific Discovery of Extraterrestrials Visiting Earth* by Courtney Brown. (Dutton, 1996).

38. *Crash at Corona: The US Military Retrieval and Cover-Up of a UFO* by Don Berliner and Stanton T. Friedman. (Marlowe & Co., 1997).
39. *Cult Science Fiction Films* by Welch Everman. (Citadel Press, 1995).
40. *The Day after Roswell: A Former Pentagon Official Reveals the US Government's Shocking UFO Cover-Up* by Philip J. Corso and William J. Birnes. (Pocket Books, 1997, 2017).
41. *Dimensions: A Casebook of Alien Contact* by Jacques Vallee. (Ballantine, 1988).
42. *E.T. 101: The Cosmic Instruction Manual for Planetary Evolution/Emergency Remedial Earth Edition* by Zoev Jho. (HarperCollins, 1995).
43. *Encounter at Bluff Ledge: A UFO Case History* by Walter N. Webb. (J. Allen Hynek Center for UFO Studies, 1994).
44. *Encounters: A Psychologist Reveals Case Studies of Abductions by Extraterrestrials* by Edith Fiore. (Doubleday, 1989).
45. *The Encyclopedia of UFOs* by Ronald D. Story. (Doubleday, 1980).
46. *Entities: Angels, Spirits, Demons and Other Alien Beings* by Joe Nickell. (Prometheus Books, 1995).
47. *Ether-Technology: A Rational Approach to Gravity Control* by Rho Sigma. (Adventures Unlimited Press, 1996).
48. *Examining the Earthlight Theory: The Yakima UFO Microcosm* by Greg Long. (J. Allen Hynek Center for UFO Studies, 1994).
49. *Extra-Terrestrial Friends and Foes* by George C. Andrews. (Illuminet Press, 1993).
50. *Extra-Terrestrials Among Us* by George C. Andrews. (Llewellyn, 1992).
51. *Extraterrestrial Archaeology: Incredible Proof We Are Not Alone* by David Hatcher Childress. (Adventures Unlimited Press, 1994).

52. *The Eyes of the Sphinx: The Newest Evidence of Extraterrestrial Contact in Ancient Egypt* by Erich Von Däniken. (Berkley, 1996).
53. *The Field Guide to Extraterrestrials* by Patrick Huyghe. (Avon, 1996).
54. *Fire in the Sky: The Walton Experience* by Travis Walton. (Marlowe & Co., 1996).
55. *Flying Saucers - Serious Business* by Frank Edwards. (Citadel Press, 1966).
56. *Flying Saucers: A Modern Myth of Things Seen in the Skies* by Carl Gustav Jung. (Princeton University Press, 1991).
57. *Forbidden Science: Journals 1957–1969* by Jacques Vallee. (Marlowe & Co., 1996).
58. *From Elsewhere: Being E. T. in America* by Scott Mandelker. (Birch Lane Press, 1995).
59. *Gifts of the Gods?: Are UFOs Alien Visitors or Psychic Phenomena?* by John Spencer. (Virgin, 1995).
60. *The Gods of Eden* by William Bramley. (Avon, 1993).
61. *The Great Science Fiction Films* by Richard Meyers. (Citadel Press, 1962).
62. *The Gulf Breeze Sightings* by Ed Walters. (Avon, 1991).
63. *A History of UFO Crashes* by Kevin D. Randle. (Avon, 1995).
64. *The Hollow Earth: The Greatest Geographical Discovery in History* by Raymond Bernard. (Citadel Press, 1991).
65. *Hollywood vs. The Aliens: The Motion Picture Industry's Participation in UDO Disinformation* by Bruce Rux. (North Atlantic Books, 1997).
66. *How To Catch a Flying Saucer* by James M. Deem. (Houghton Mifflin, 1991).
67. *How To Defend Yourself Against Alien Abduction* by Ann Druffel. (Three Rivers Press, 1998).
68. *The Hynek UFO Report* by J. Allen Hynek. (Dell, 1997).

69. *Incident at Exeter* and *Interrupted Journey* by John G. Fuller. (Fine Communication, 1996).
70. *Inside Heaven's Gate: The UFO Cult Leaders Tell Their Story in Their Own Words* by Hayden Hewes and Brad Steiger. (Signet, 1997).
71. *Intruders: The Incredible Visitations at Copley Woods* by Budd Hopkins. (Random House, 1987).
72. *Left at East Gate: A First-Hand Account of the Bentwaters-Woodbridge UFO Incident, Its Cover-Up, and Investigation* by Peter Robbins and Larry Warren. (Marlowe & Co., 1997).
73. *Light Years: An Investigation into the Extraterrestrial Experiences of Eduard Meier* by Gary Kinder. (Atlantic Monthly Press, 1987).
74. *Majestic* by Whitley Streiber. (Berkley, 1990).
75. *Making Contact: A Serious Handbook for Locating and Communicating with Extraterrestrials* by Bill Fawcett. (Morrow, 1997).
76. *The Meeting of Science and Spirit: Guidelines for a New Age* by John White. (Paragon House, 1990).
77. *The Monuments of Mars: A City on the Edge of Forever* by Richard C. Hoagland. (Frog, Ltd., 1996).
78. *Moongate: Suppressed Findings of the US Space Program* by William L. Brian II. (Future Science Research Publishing, 1982).
79. *Mysteries of the Unexplained.* (Reader's Digest, 1982).
80. *Our Cosmic Ancestors* by Maurice Chatelain. (Temple Golden Publications, 1988).
81. *Out There: The Government's Secret Quest for Extraterrestrials* by Howard Blum. (Pocket, 1991).
82. *Outer Space Connection* by Alan Landsburg and Sally Landsburg. (Bantam, 1975).
83. *Project Blue Book: The Top Secret UFO Findings Revealed* by Brad Steiger. (Ballantine, 1995).

84. *Project Mindshift: The Re-education of the American Public Concerning Extraterrestrial Life 1947-Present* by Michael Mannion. (M. Evans & Co., 1998).
85. *Psychic Sasquatch and the UFO Connection* by Jack Lapseritis. (Blue Water Publisher, 1998).
86. *The Randle Report: UFOs in the '90s* by Kevin D. Randle. (M. Evans & Co., 1997).
87. *Report on Communion: An Independent Investigation of and Commentary on Whitley Streiber's Communion* by Ed Conroy. (Morrow, 1989).
88. *The Science in Science Fiction* by Peter Nicholls. (Knopf, 1983).
89. *Scientific Study of Unidentified Flying Objects* by Dr. Edward U. Condon. (Bantam, 1969).
90. *Silent Invasion: The Shocking Discoveries of a UFO Researcher* by Ellen Crystall. (St. Martin's, 1991).
91. *Transformation: The Breakthrough* by Whitley Streiber. (Avon, 1988).
92. *The Truth About the UFO Crash at Roswell* by Kevin D. Randle and Donald R. Schmitt. (Evans, 1994).
93. *UFO Abductions in Gulf Breeze* by Ed Walters and Frances Walters. (Avon, 1994).
94. *UFO Abductions: A Dangerous Game* by Philip J. Klass. (Prometheus Books, 1989).
95. *The UFO Casebook* by Kevin D. Randle. (Warner, 1989).
96. *UFO Chronicles of the Soviet Union: A Cosmic Samizdat* by Jacques Vallee. (Ballantine, 1992).
97. *The UFO Cover-up: What the Government Won't Say* by Lawrence Fawcett and Barry Greenwood. (Simon & Schuster, 1993).
98. *UFO Crash at Roswell* by Kevin D. Randle and Donald R. Schmitt. (Avon, 1991).

99. *UFO Encounters & Beyond* by Jerome Clark. (NAL, 1993).

100. *The UFO Encyclopedia* by Jerome Clark. (Omnigraphics, 1990-1996).

    Volume 1: *UFOs in the 1980s* (1990).

    Volume 2: *The Emergence of a Phenomenon: UFOs from the Beginning through 1979* (1992).

    Volume 3: *High Strangeness: UFOs from 1960 through 1979* (1996).

101. *The UFO Encyclopedia* by John Spencer. (Avon, 1993).

102. *The UFO Enigma: The Definitive Explanation of the UFO Phenomenon* by Donald H. Menzel and Ernest H. Taves. (Doubleday, 1997).

103. *The UFO Evidence*, Richard Hall, editor. (Barnes & Noble Books, 1997).

104. *The UFO Experience: A Scientific Inquiry* by J. Allen Hynek. (Regnery, 1972).

105. *The UFO Guidebook* by Norman J. Briazack and Simon Mennick. (Citadel Press, 1978).

106. *The UFO Handbook: A Guide to Investigating, Evaluating and Reporting UFO Sightings* by Allan Hendry. (Doubleday, 1979).

107. *The UFO Invasion: The Roswell Incident, Alien Abductions, and Government Cover-ups* by Kendrick Frazier, editor, et al. (Prometheus, 1997).

108. *The UFO Phenomenon*. Time-Life Books, 1997.

109. *UFO Quest: In Search of the Mystery Machines* by Alan Watts. (Blandford, 1994).

110. *The UFO Report* by Timothy Good, editor. (Avon, 1991).

111. *UFO Retrievals: The Recovery of Alien Spacecraft* by Jenny Randles. (Sterling, 1995).

112. *The UFO Verdict: Examining the Evidence* by Robert Sheaffer. (Prometheus, 19981).

113. *UFO Visitation: Preparing for the Twenty-First Century* by Alan Watts. (Blandford, 1996).
114. *UFO: A Deadly Concealment: The Official Cover-Up* by Derek Sheffield. (Blandford, 1996).
115. *UFO: The Complete Sightings* by Peter Brookesmith. (Barnes & Noble Books, 1995).
116. *UFO: The Definitive Guide to Unidentified Flying Objects and Related Phenomena* by David Ritchie. (Facts on File Communications, 1994).
117. *UFO: The Government Files* by Peter Brookesmith. (Barnes & Noble Books, 1996).
118. *UFOs and How To See Them* by Jenny Randles. (Sterling, 1993).
119. *UFOs are Real: Here's the Proof* by Edward Walters. (Avon, 1997).
120. *UFOs: A Manual for the Millenium* by Phil Cousineau. (Harper San Francisco, 1995).
121. *UFOs: Psychic Close Encounters: The Electromagnetic Indictment* by Albert Budden. (Blandford, 1995).
122. *UFOs: The Secret History* by Michael Heseman. (Marlowe & Co., 1998).
123. *UFOs: The Sightings of Alien People and Spacecraft from the Earliest Centuries to the Present Day* by Robert Jackson. (Smithmark, 1992).
124. *Unexplained Mysteries of the 20th Century* by Janet Bord and Colin Bord. (Contemporary Books, 1989).
125. *The Watchers II: Exploring UFOs and the Near-Death Experience* by Raymond E. Fowler. (Wild Flower Press, 1995).
126. *The Watchers: The Secret Design Behind UFO Abduction* by Raymond E. Fowler. (Bantam, 1991).
127. *We Are Not Alone: The Continuing Search for Extraterrestrial Intelligence* by Walter Sullivan. (Plume, 1994).

128. *We Discovered Alien Bases on the Moon II* by Fred Steckling. (G.A.F. International, 1997).
129. *World Atlas of UFOs: Sightings, Abductions and Close Encounters* by John Spencer. (Smithmark, 1992).

# 103 Novels about UFOs and Aliens

Many of these novels are classics in the field; some of you will have read a great many titles on this list.

Also, some of these novels are undoubtedly out of print, but many local public libraries will have some of these older titles on their shelves, and if a title or author intrigues you, it would be worth checking out your local library. Interlibrary loan offers you the opportunity to check out books from just about any library in your state, so even if your hometown library doesn't have a book you're looking for, you still may be able to get a copy through the ILL system.

For all of these novels, we have provided the title, author(s), and year of publication. We have deliberately omitted the publisher in order to save space and also because hardcovers and softcovers were usually a different imprint (even within the same publishing house), and you don't really care who published it anyway. If the book's in print, any bookstore or online book service (web favorites are, of course, www.amazon.com and www.barnesandnoble.com) will be able to get it for you. If it isn't, then one of the used book services (especially www.abe.com) will find it in any edition and from any publisher; and if you look for it in a library, the title and author are all you will ever need to search the stacks. Thus, no publishers.

There are also some oddities included on this list, such as the 1970 "sex-on-an-alien-planet classic," *Runts of 61 Cygni C* by James Grazier, which science fiction critic David Pringle described as "one of the prime contenders for the title of Worst SF Novel Ever Published." That review *alone* might make you want to seek it out.

A good way to dip into this 103-title list would be to track down some of the books by your favorite sci-fi authors; that should get you started on a wondrous journey!

1. The *Alien* Series by Alan Dean Foster (1979–1986)
2. *The Andromeda Strain* by Michael Crichton (1969)
3. *Battlefield Earth* by L. Ron Hubbard (1982)
4. *Between Planets* by Robert A. Heinlein (1951)
5. *Beyond the Blue Event Horizon* by Frederik Pohl (1980)
6. *The Body Snatchers* by Jack Finney (1955)
7. *A Case of Conscience* by James Blish (1958)
8. *Chapterhouse: Dune* by Frank Herbert (1985)
9. *Childhood's End* by Arthur C. Clarke (1963)
10. *Children of Dune* by Frank Herbert (1976)
11. *Close Encounters of the Third Kind* by Steven Spielberg (1977)
12. *Contact: A Novel* by Carl Sagan (1985)
13. *The Cosmic Puppets* by Philip K. Dick (1957)
14. *Cradle* by Arthur C. Clarke, Gentry Lee (1988)
15. *The Currents of Space* by Isaac Asimov (1952)
16. *A Day for Damnation* by David Gerrold (1984)
17. *The Day of the Triffids* by John Wyndham (1951)
18. *The Day the Martians Came* by Frederik Pohl (1988)
19. *The Dispossessed* by Ursula K. LeGuin (1974)
20. *The Divine Invasion* by Philip K. Dick (1982)
21. *Do Androids Dream of Electric Sheep?* Philip K. Dick (1968)
22. *The Dosadi Experiment* by Frank Herbert (1977)
23. *Double Star* by Robert A. Heinlein (1956)
24. *Dune* by Frank Herbert (1965)
25. *Dune Messiah* by Frank Herbert (1969)
26. *E.T: The Extra-Terrestrial* by William Kozwinkle (1982)

27. *Earthlight* by Arthur C. Clarke (1955)
28. *Ender's Game* by Orson Scott Card (1985)
29. *Enemy Mine* by David Gerrold, Barry B. Longyear (1985)
30. *The Enemy Within* by L. Ron Hubbard (1986)
31. *The First Men in the Moon* by H. G. Wells (1901)
32. *Foundation and Earth* by Isaac Asimov (1986)
33. *Foundation and Empire* by Isaac Asimov (1952)
34. *Foundation* by Isaac Asimov (1951)
35. *Foundation's Edge* by Isaac Asimov (1982)
36. *Fountains of Paradise* by Arthur C. Clarke (1979)
37. *Gateway* by Frederik Pohl (1977)
38. *The God-Emperor of Dune* by Frank Herbert (1981)
39. *Gods of Riverworld* by Philip José Farmer (1983)
40. *The Gods Themselves* by Isaac Asimov (1972)
41. *Have Space-Suit—Will Travel* by Robert A. Heinlein (1958)
42. The *Helliconia* Series by Brian W. Aldiss (1982-1985)
43. *Heretics of Dune* by Frank Herbert (1984)
44. *The Hitch-Hiker's Guide to the Galaxy* by Douglas Adams (1979)
45. *Insomnia* by Stephen King (1994)
46. *The Invaders Plan* by L. Ron Hubbard (1985)
47. *It* by Stephen King (1986)
48. *Jonathan Strange and Mr. Norrell* by Suzanne Clarke (2004)
49. *The Left Hand of Darkness* by Ursula K. Le Guin (1969)
50. *Life, the Universe and Everything* by Douglas Adams (1982)
51. *Lord of Light* by Roger Zelazny (1967)
52. *Lord Valentine's Castle* by Robert Silverberg (1980)
53. *Majestic* by Whitley Strieber (1989)
54. *The Making of the Representative for Planet 8* by Doris Lessing (1982)
55. *The Man Who Fell to Earth* by Walter Tevis (1963)

56. *The Marriages Between Zones Three, Four, and Five* by Doris Lessing (1980)
57. *The Mote in God's Eye* by Jerry Pournelle, Larry Niven (1974)
58. *Nemesis* by Isaac Asimov (1989)
59. *Neuromancer* by William Gibson (1984)
60. *Nightmare Journey* by Dean R. Koontz (1975)
61. *Out of the Silent Planet* by C. S. Lewis (1938)
62. The *Passage* Series by Justin Cronin (2010, 2012, 2016)
63. *Pebble in the Sky* by Isaac Asimov (1950)
64. *Planet of Exile* by Ursula K. Le Guin (1966)
65. *Prelude to Foundation* by Isaac Asimov (1988)
66. *Princess of Mars* by Edgar Rice Burroughs (1917)
67. *The Puppet Masters* by Robert A. Heinlein (1951)
68. *Red Planet* by Robert A. Heinlein (1949)
69. *Rendezvous with Rama* by Arthur C. Clarke (1973)
70. *The Restaurant at the End of the Universe* by Douglas Adams (1980)
71. *Riders of the Purple Wage* by Philip Jose Farmer (1992)
72. *Ringworld* by Larry Niven (1970)
73. *The Ringworld Engineers* by Larry Niven (1980)
74. *Runts of 61 Cygni C* by James Grazier (1970)
75. *Schrodinger's Cat: The Universe Next Door* by Robert Anton Wilson (1979)
76. *Second Foundation* by Isaac Asimov (1952)
77. *The Sentimental Agents in the Volyen Empire* by Doris Lessing (1983)
78. *Shikasta* by Doris Lessing (1979)
79. *The Sirian Experiments* by Doris Lessing (1981)
80. *Solaris* by Stanislaw Lem (1961)
81. *So Long, and Thanks for All the Fish* by Douglas Adams (1984)

82. *The Space Vampires* by Colin Wilson (1977)
83. *Sphere* by Michael Crichton (1987)
84. *The Star Beast* by Robert A. Heinlein (1954)
85. *The Star King* by Jack Vance (1964)
86. *Stars in My Pocket Like Grains of Sand* by Samuel R. Delaney (1984)
87. *The Stars My Destination* by Alfred Bester (1956)
88. *Starship Troopers* by Robert A. Heinlein (1959)
89. *Stranger in a Strange Land* by Robert A. Heinlein (1961)
90. *Titan* by John Varley (1979)
91. *The Tommyknockers* by Stephen King (1987)
92. *2001: A Space Odyssey* by Arthur C. Clarke (1968)
93. *2061: Odyssey Three* by Arthur C. Clarke (1987)
94. *2010: Odyssey Two* by Arthur C. Clarke (1982)
95. *2312* by Kim Stanley Robinson (2012)
96. *Up the Walls of the World* by James Tiptree Jr. (1978)
97. The *Voyagers* Series by Ben Bova (1981-1987)
98. *War of the Worlds* by H. G. Wells (1898)
99. *The Weapon Shops of Isher* by A. E. van Vogt (1951)
100. *Who Goes There?* by John W. Campbell Jr. (1948)
101. *Wizard* by John Varley (1980)
102. *The Word for World is Forest* by Ursula K. Le Guin (1976)
103. *Wyrms* by Orson Scott Card (1987)

Also . . .

The *Star Wars* novel series

The *Star Trek* novel series

# UFO or Extraterrestrial-Themed TV Shows

Science Fiction is so popular it even has its own TV channel, The SyFy Channel. And science fiction TV shows have been a staple in TV programming since the earliest days of the invention.

This feature looks at sixty-seven out-of-this-world TV shows that used UFOs and/or aliens in their plots.

### The Adventures of Superman

Superman came to earth from another planet in a spaceship, so that makes him an ET! (Albeit one who can run *really* fast, and bend steel in his bare hands, and leap tall buildings in a single bound, and who, disguised as . . . well, you know the rest!) (July 1951–November 1957—104 episodes; syndicated and network daytime).

### Alf

The alien Alf's (*A*lien *L*ife *F*orm) spaceship crashed into the Tanners' garage, and the family hid him in their kitchen (at least until the show was canceled!) (September 22, 1986–June 18, 1990; NBC).

### Alien Nation

A slave transport spaceship from the planet Tencton on its way to another planet crashes in the Mojave Desert, and the freed aliens decide to make Earth their home. Everyone on Earth knows about the Newcomers, and the aliens do their best to try and fit in among their new neighbors. The hard-line Purists are against the aliens and want them to leave. The Newcomers (who have larger heads and mottled pigmentation instead of hair) face bigotry and racism on a daily basis as they try to make a life on their adopted home. This series looked at racism and intolerance much the way the original *Star Trek* did: by using extraterrestrial characters to make the point. Hell, even the title of this series bespoke sociocultural isolation: *alienation*! (September 18, 1989–July 26, 1991; Fox).

### Amazing Stories

This science fiction/fantasy anthology was one of Steven Spielberg's infrequent TV projects and, in all honesty, it was probably too good for TV. *Amazing Stories* featured weekly adaptations of terrific sci-fi and fantasy tales. Spielberg's standing in Hollywood lured people to the show who would usually never go *near* TV, including (as directors) Martin Scorsese, Clint Eastwood, Burt Reynolds, Paul Bartel, and Spielberg himself; as well as such actors as Kevin Costner, Drew Barrymore, Charlie Sheen, and Sam Waterston. (September 29, 1985–May 15, 1987; NBC).

### Babylon 5

This was a drama set on a space station in the year 2258 where beings from many planets lived and worked together. Part sci-fi adventure series; part space soap opera, *Babylon 5* soon found a devoted audience which blossomed into a cult following within the first two seasons. (1992–Syndicated).

### Battlestar Galactica

This elaborately produced and groundbreakingly expensive ($1 million per hour episode) series set thousands of years in the future was essentially "*Star Wars* for TV." In fact, *Battlestar Galactica* was *so* much like *Star Wars* that ABC was sued by George Lucas and company for ripping off their film! The show was magnificent for television, though, boasting special effects usually only found in movies, and spaceships built by John Dykstra, who (not so coincidentally) worked on *Star Wars.* Set in the Seventh Millenium A.D., Galactica, the only surviving battlestar after a devastating rampage by the Cylons, tried to get back to Earth while being pursued by the Cylons. A second *Galactica* series—Galactica 1980—aired for one season following the original *Battlestar Galactica.* (September 17, 1978–August 17, 1980; ABC; December 2003–March 20, 2009; Sci-Fi Channel.).

### Battlestar Galactica (2004)

Extensively reimagined from the eighties original, Ronald D. Moore and David Eick's *Battlestar Galactica* tracks the ongoing conflict between humans and Cylons, a race of man-made cyborgs that rebel against their creators, and in an unprecedented sneak attack, lay waste to the colonies. A single military vessel, the eponymous *Battlestar Galactica*, manages to escape with a freight of the last humans in the universe to the mythical planet Earth. (October 18, 2004–March 20, 2009; Syfy).

### Buck Rogers and Buck Rogers in the 25th Century

*Buck Rogers* was originally a twelve-chapter Saturday matinée theatrical series starring Buster Crabbe as a pilot who spends five hundred years in suspended animation only to wake up in a different century where the world has been conquered and ruled by a villain called Killer Kane. The series migrated to television. Buck Rogers was responsible for the safety of the universe. Yikes. How's that for a job description? Buck used special scientific devices and gadgets to foil the never-ending parade of galactic villains. *Buck Rogers in the 25th Century* aired for three seasons on NBC thirty years later and boasted better special effects (the original show's effects were genuinely cheesy—although not *Plan 9 From Outer Space* cheesy!) and this 1980s version was more successful due to the increased interest in science fiction in general, thanks to the popularity of *Star Wars* and other genre movies. One cool element of the remake was a character named Admiral Asimov: He was a direct descendant of the legendary writer Isaac Asimov! (There was also a robot named "Crichton.") (Original: April 15, 1950–January 30, 1951; ABC; Remake: September 20, 1979–April 16, 1981; NBC).

### Caprica

*Caprica* is a *Battlestar Galactica* spin-off prequel that focuses on the origin of the Cylon race fifty-eight years before the explosive first episode of the reimagined *Battlestar Galactica.* (January 22–November 30, 2010; SyFy).

### Captain Video and His Video Rangers

This 1950s children's show can rightfully be called the granddaddy of all science fiction TV shows. The special effects were almost nonexistent (the show had a prop budget of $25 a week); the sets were something an inventive twelve-year-old could put together in his garage, and the stories were shamelessly melodramatic. And yet *Captain Video* was a mega-hit for its time. The show was such an icon in the early fifties that *The Honeymooners* did an entire episode around Ed Norton (Art Carney) not wanting to miss an episode of the show. (Norton was one of the Captain's stalwart at-home "Video Rangers!") The show's writer, Maurice Brock, consistently came up with new futuristic weapons and devices for the show (The Opticon Scillometer, Atomic Rifle, and Trisonic Compensator were just a few of his "inventions."). He always tried to work a moral into each episode, foreshadowing the era of TV shows like *Leave It To Beaver, Father Knows Best*, and *The Adventures of Ozzie and Harriet*—shows that did the same thing with almost every episode—but boasted a somewhat larger budget! *Captain Video and His Video Rangers* is included here because the Captain frequently battled with evil aliens, including Mook the Moon Man and Kul of Eos. The show offered decoder rings and copies of the Captain's helmet to its loyal viewers, and even enjoyed a brief run as a theatrical serial. (June 27, 1949–April 1, 1955; Dumont).

### Defiance

Another science fiction western drama television series, this show is of a far more serious bent that Whedon's *Firefly*. The show takes place on a post-apocalyptic future Earth that is almost unrecognizable as our own, having been terraformed by alien Votan technology. The series follows Sheriff Joshua Nolan as he polices the town of Defiance, a mining community where human and alien species coexist. (April 15, 2013–August 28, 2015; SyFy).

### Doctor Who

A much-beloved and culturally-significant science fiction series, *Doctor Who* follows the adventures of a shape-shifting, immortal Time Lord across time and space as he seeks to help all those in need and defend the universe. (Classic series: November 23, 1963–December 6, 1989 / Revived series: March 26, 2005–present; BBC).

### Earth: Final Conflict

*Earth: Final Conflict* is a Canadian science fiction television series based on notes kept by Gene Roddenberry's widow, Majel Barrett-Roddenberry, that provided the concept for the series. It ran for five seasons between October 6, 1997, and May 20, 2002. (October 6, 1997–May 20, 2002; syndication, CTV, The New Net).

### Eureka

*Eureka* is a Syfy television show about the fictional town of Eureka, Oregon inhabited almost entirely by geniuses. Each episode features a bit of technology gone wrong that town sheriff, Jack Carter, must make right, with a little help from the local talent. (July 18, 2006–July 16, 2012; Syfy).

### The Event

*The Event* is an eclectic mix of science fiction, action/adventure, and political allegory centering on a group of extraterrestrials, some of whom have been detained by the United States government for sixty-six years since their crash-landing in Alaska, and some of whom have assimilated to human society. (September 20, 2010–May 23, 2011; NBC).

### Extant

Cancelled after only two seasons, *Extant* tells the story of astronaut Molly Woods (played by Halle Berry), who returns to Earth from a solo space mission pregnant. A *solo* mission. Hmm. Steven Spielberg served as executive producer. (July 9, 2014–September 9, 2015; CBS).

### Falling Skies

Beginning six months after a global invasion of Earth, *Falling Skies* is yet another series presided over by executive producer Steven Spielberg. It revolves around a Boston University professor named Tom Mason who spearheads a resistance force dubbed the 2nd Massachusetts Militia Regiment while also on a quest to find his son. (June 19, 2011–August 30, 2015; TNT).

### Farscape

Astronaut John Crichton's luck runs out quite abruptly one day when, on an experimental flight, he is sucked into a wormhole. Determined to return to Earth, he must contend with a corrupt militaristic organization known as the Peacekeepers and the deadly secret locked in his own mind! (March 19, 1999–March 21, 2003; Nine Network, SyFy).

### Firefly

A science fiction space Western set in the year 2517, centering on the renegade crew of the "Firefly-class" spaceship, Serenity, and their various adventures across a galaxy that is equal parts *Star Wars* and Clint Eastwood. Joss Whedon pitched the show as "nine people looking into the blackness of space and seeing nine different things," and that rich diversity of character is central to the show's appeal. It's a rollicking good time that, regrettably, never received the second season it so rightfully deserved. This series only lasted one season and fourteen episodes but has become one of the most beloved cult TV series of all time. There has been a consistent effort on the part of fans to convince *someone* to bring the show back. The nine crewmembers make their living from smuggling and other sundry rogue doings while trying to evade the Reavers, a race of humans that has devolved into a barbaric, cannibalistic horde of animal-like invaders. Over the past fifteen years, the *Firefly* franchise has survived with a variety of products for fans, including a movie (*Serenity*) comic books, and a video game. (Not to mention T-shirts and other *Firefly* memorabilia.) (September 20–December 20, 2002; Fox)

### First Wave

A Canadian science fiction drama television series created by Chris Brancato, who co-wrote an early version of the script for the *X-Files* episode "Eve," and executive produced by Francis Ford Coppola, the show has some incredible credentials, to say the least. *First Wave* follows Kincaid Lawrence "Caid" Foster's battle against a race of aliens called the Gua who plan to enslave humankind and live among us in the meanwhile, biding their time. (September 9, 1998–February 7, 2001; Space, Syfy).

### The 4400

Pronounced "the forty-four hundred," *The 4400* is a CBS Paramount Network Television program about a group of 4,400 people who suddenly appear in the Cascade Range foothills near Mount Rainier, Washington with no memory of how they got there, or why. As the story unfolds, it's revealed that each of these people vanished in beams of white light at one time or another—some as far back as 1938! (July 11, 2004–September 16, 2007; USA Network).

### Fringe

Superb TV series about the fictional Fringe Division of the FBI, a clandestine branch that investigates paranormal, and often horrifying "fringe" events, including parallel universes, time travel, alternate timelines, and teleportation. Variously considered an update or a riff on sci-fi staple *The X-Files*, *Fringe* follows Special Agent Olivia Dunham and Walter and Peter Bishop, members of the FBI's fictional division, as they investigate paranormal, supernatural, and inexplicable phenomena out of their lab at Harvard University. And it even had a "mad scientist" in the character of Dr. Walter Bishop (played by John Noble). It lasted five seasons and 100 episodes. (September 9, 2008–January 18, 2013; Fox.)

### Gilligan's Island

UFOs? Extraterrestrials? *Gilligan's Island??* Yup, because a few episodes of this astonishingly popular series had sci-fi/alien themes. In "Smile, You're on Mars Camera," a camera headed for Mars lands on the island and the scientists observing think the Castaways are Martians; in "Nyet, Nyet - Not Yet," a Russian space capsule lands on the island; in "Splashdown," the Castaways try to contact an orbiting spacecraft; and in "Meet the Meteor," a strange meteor crashes on the island and gives off cosmic rays that will cause all the Castaways to die of old age in a week. (September 26, 1964–September 4, 1967; CBS).

### The Invaders

This late sixties series produced by Quinn Martin centered around architect David Vincent (Roy Thinnes) and his efforts to convince the world that the aliens had landed, that they had infiltrated humanity, and that they were an advance guard preparing the way for a full-scale alien invasion. (January 10, 1967–September 17, 1968; ABC).

### Land of the Giants

This sci-fi fantasy produced by Irwin Allen (*Voyage to the Bottom of the Sea, The Towering Inferno*) was about seven 1980s Earthlings who were accidentally drawn into a space warp and transported to an Earth-like planet where the locals were approximately twelve times as large as they were. As the hapless seven tried to survive in this bizarre place, Inspector Kobrick (an homage to Stanley Kubrick, perhaps?) of the clandestine security agency the S.I.B, tried to locate them and return them all safely to Earth. (September 22, 1968–September 6, 1970; ABC).

### Land of the Lost

*Land of the Lost* details the adventures of Rick, Will, and Holly Marshall, who are trapped in an alternate universe inhabited by dinosaurs. (September 7, 1974–December 4, 1976; NBC).

### Lois and Clark: The New Adventures of Superman

More fun with the superhero from Krypton and his best girl! (September 12, 1993–June 14, 1997; ABC).

### Lost

A drama with science fiction and paranormal elements, *Lost* tells the story of the survivors of a plane crash stranded on a South Pacific island. Who are "The Others?" What the hell is the smoke monster? It lasted six seasons and many fans were disappointed that it concluded with still unanswered questions. (September 22, 2004–May 23, 2010; ABC).

### Lost in Space

The space family Robinson travels from planet to planet, meeting extraterrestrials and other perils, trying desperately to protect themselves from the evil Dr. Smith and return to Earth safely. Incredibly, cheesy special effects did not lessen the enthusiasm fans felt for this series and, ironically, incredible special effects in the 1998 movie adaptation of the show did not win over original *Lost in Space* fans, nor bring new ones into the franchise. (September 15, 1985–September 11, 1968; CBS).

### Mork & Mindy

In the tradition of Gore Vidal's *Visit to a Small Planet*, the alien Mork from Ork lands on Earth, befriends the comely Mindy, and wreaks hilarious havoc on his nearest and dearest's lives for five seasons. *Mork & Mindy* was the late Robin Williams's breakout hit. (September 14, 1978–June 10, 1982; ABC).

### My Favorite Martian

Tim O'Hara came upon a Martian in a crashed spaceship, nursed him back to health, and passed him off as his Uncle Martin for three successful seasons. Uncle Martin had many "Martian" powers, the coolest of which was being able to move objects simply by pointing at them. (September 29, 1963–September 4, 1966; CBS).

### The Neighbors

*The Neighbors* is an American television science fiction sitcom about a human family living in a neighborhood of other alien families. (September 26, 2012–April 11, 2014; ABC).

### The 100

The 100 (pronounced "the hundred") is a post-apocalyptic science fiction drama based very loosely on a 2013 novel by Kass Morgan. The series follows a group of teens returning to Earth after a nuclear apocalypse. (March 19, 2014–present; The CW).

### Orphan Black

Canadian science fiction series about several people who are all clones. (March 30, 2013–present; BBC America, SPACE).

### The Outer Limits (Original)

You know a TV show has achieved a rare level of pop culture icon status when its title and opening monologue are now an immediately recognized part of TV's heritage. The *Superman* intro and Rod Serling's *Twilight Zone* intros are two examples of this kind of influence. The intro to *The Outer Limits* is another: *"There is nothing wrong with your television set. Do not attempt to adjust the picture. We are controlling transmission. We will control the horizontal. We will control the vertical. We can change the focus to a soft blur—or sharpen it to crystal clarity. For the next hour, sit quietly and we will control all that you see and hear. You are about to participate in a great adventure. You are about to experience the awe and mystery which reaches from the inner mind to The OUTER LIMITS." The Outer Limits* was one of those cultural touchstones that immediately engaged viewers and found a devoted audience. The Showtime Cable Network resurrected the title and produced all-new episodes (in color this time) of the series beginning in 1995—still retaining the focus on aliens, UFOs, paranormal phenomena, and just all-around general weirdness. (September 16, 1963–January 16, 1965; ABC).

### Project UFO

This was the only UFO TV show that was actually based on government research, specifically Project Blue Book. Executive Producer Jack Webb used the unexplained Project Blue Book sightings to create dramatizations for this series, which involved two Air Force investigators who traveled around the country interviewing the people who had reported the sightings the government could not explain. (February 19, 1978–August 30, 1979; NBC).

### Quark

This short-lived sci-fi sitcom was about a space station garbage collector named Adam Quark who often found himself inadvertently caught up in adventures involving alien beings like Zorgon the Malevolent and the evil High Gorgon. Quark's boss was The Head (shades of *3rd Rock*!), and his co-workers included a human vegetable, a First Officer who was both male and female, and a robot. *Quark* was created as a parody of (and to cash in on) space opera adventures such as the *Star Wars* series, which was a huge hit at the time. (February 24, 1978–April 14, 1978; NBC).

### Rick and Morty

*Rick and Morty* is a science fiction comedy that revolves around Rick, a genius scientist with bad habits, and his grandson Morty, a wimp with a good heart. The show began as a *Back to the Future* parody but quickly developed into its own beast, tackling and subverting every sci-fi trope around. As brainy as often as it can be crude, *Rick and Morty* is nevertheless a shining example of how science fiction doesn't need to take itself very seriously to be entertaining. (December 2, 2013–present; Adult Swim).

### Roswell

In Roswell, New Mexico, three teenaged human/alien hybrids named Max Evans, Isabel Evans, and Michael Guerin quietly await the day they can return to their home planet and save their dying race. This popular show was a gateway to science fiction for many a teenager through the late 90s and early 00s. (October 6, 1999–May 14, 2002; The WB, CPN, Syfy).

### Smallville

Among the longest-running American sci-fi/fantasy television shows ever, *Smallville* follows Superman, everyone's favorite alien superhero in his early years in Smallville, Kansas, from high school to his career with the Daily Planet and beyond, encountering familiar faces—friend *and* foe—along the way. (October 16, 2001–May 13, 2011; The WB [2001–2006]/The CW [2006–2011]).

### Space: Above and Beyond

Ranked 50th in IGN's top 50 Sci-Fi TV Shows and described as "yet another sci-fi show that went before its time," *Space: Above and Beyond* is set in the years 2063–2064 and focuses on the "Wildcards," members of the United States Marine Corps Space Aviator Cavalry, 58th Squadron. (September 24, 1995–June 2 1996; Fox).

### Space: 1999

The Moon is blasted free of its orbit when stored nuclear wastes accidentally detonate, and suddenly, the crew of Moonbase Alpha is hurtling into deep space, the Moon itself their out-of-control spaceship. They run into aliens, of course, and have interstellar adventures galore, but all of the big-budget special effects couldn't prolong the show's survival, which starred the husband and wife team of Martin Landau and Barbara Bain. (1974–1976 – 48 episodes; Syndicated).

### Stargate SG-1

One of the most successful TV spin-offs ever, based on the 1994 hit movie of the same name, *Stargate SG-1* reintroduces Jonathan "Jack" O'Neill and Daniel Jackson alongside new characters such as Teal'c, George Hammond, and Samantha Carter. The first season deals with a military-science expedition team diskovering how to use the ancient device known as the Stargate and their battle against the sinister forces at large in the universe. (July 27, 1997–March 13, 2007; Showtime, Syfy).

### Starman

A one-season series based on the hit 1984 movie about an alien who lands on Earth and falls in love with an Earth woman. In this series, he returns to Earth to help raise the son he fathered during his liaison with the woman played by Karen Allen in the movie. (September 19, 1986–September 4, 1987; ABC).

### Star Trek: The Original Series

The classic sci-fi series—and a show (and movie series!) more popular now than ever. Need we say more? The original series was purchased from Gene Roddenberry by Desi Arnaz for Desilu. After Desilu was purchased by Paramount, Paramount became the owner of the *Star Trek* franchise and is still producing it today. (September 8, 1966–September 2, 1969; NBC).

### Star Trek: Deep Space Nine

A spin-off from *Star Trek: The Next Generation* that was set on a space station and introduced us to the Ferengi and the Borg. *Deep Space Nine* seems to be the least popular of the *Star Trek* spin-offs. (1992–Syndicated).

### Star Trek: Enterprise

A prequel to *Star Trek: The Original Series*, focusing mainly on the Temporal; Cold War. (September 26, 2001–May 13, 2005; UPN).

### Star Trek: The Next Generation

The first of three (so far) TV spin-offs of the original *Star Trek* series, *ST: TNG* is overwhelmingly considered by fans to be the best of the bunch. The character of Captain Jean-Luc Picard (Patrick Stewart) is as brilliant a creation as was Captain James T. Kirk, and the interior of the new Enterprise was eight times as large as the original starship. In 1994, as the series ended, the well-received movie *Star Trek: Generations* came out, which, in terms of story arc, actually served as a bridge between the original series and the *Next Generation* series—even though *TNG* was no longer producing new episodes. (1987–1994 – 178 episodes; Syndicated).

### Star Trek: Voyager

This third (very popular) spin-off of *Star Trek* returned to the show's roots by setting the action on a starship. Voyager was commanded by Captain Kathryn Janeway and manned by a holographic physician named Doc, a Native American, a half-Klingon, a Romulan, and many other assorted terrestrials and extraterrestrials. (January 16, 1995–; UPN).

### Star Wars

*The Clone Wars:* A CGI animated television series created by George Lucas himself that fills in the gaps between *Attack of the Clones* and *Revenge of the Sith*. A must for fans of the franchise! (October 3, 2008–March 7, 2014; Cartoon Network).

### The Strain

Guillermo del Toro's horror/vampire series, seasoned with a touch of zombie. This four-season series was based on Guillermo del Toro and Chuck Hogan's novel trilogy of the same name. (July 13, 2014–present; FX.)

### Stranger Things

Based in part upon the stories of the 'Montauk Boys,' in which the US government brought children to a secret facility at Montauk Long Island for sensory enhancement, Netflix's breakout science fiction smash-hit follows the disappearance of a young boy in Hawkins, Indiana, and his friends' quest to find him. Along the way, they encounter a telekinetic girl known only as Eleven; a shady government organization; parallel dimensions; and other elements borrowed and reupholstered from eighties sci-fi pop culture, including the works of Steven Spielberg, Stephen King, George Lucas, and John Carpenter. Pure nostalgia! (July 15, 2016–present; Netflix).

### Superboy: Clark Kent

The College Years! (1988–1991 – 78 episodes; Syndicated).

### 3rd Rock From The Sun

An over-the-top, yet wonderfully written and acted sitcom about a four-man (four-being?) alien crew who are assigned to study Earth by acting like humans. Hilarity ensues. (NBC).

### Terminator: The Sarah Connor Chronicles

A spin-off from the *Terminator* movies that begins four years after the end of *Terminator 2*. (January 13, 2008–April 10, 2009; Fox.)

### Tom Corbett: Space Cadet

This early fifties children's show was produced by CBS in an attempt to capitalize on the enormous success of the DuMont network's "golden child," *Captain Video and His Video Rangers*. *Tom Corbett* had a bigger budget than *Captain Video* and boasted better special effects. The show was about the adventures of Tom Corbett, a cadet at the Space Academy and his partners, the Earther Roger and the Venusian Astro. The series was based on the novel *Space Cadet* by legendary sci-fi author Robert Heinlein and was extremely popular during its two-year run. (October 2, 1950–September 26, 1952; CBS, ABC, NBC).

### Torchwood

*Torchwood* is a British science fiction television spin-off from the 2005 revival of *Doctor Who* whose themes of existentialism, sexuality, and morality are aimed at a decidedly more mature audience. It follows a team of alien hunters based out of the fictional Torchwood Institute, led by Captain Jack Harkness, an immortal ex-con from the distant future. (October 22, 2006–September 15, 2011; BBC, Starz).

### Threshold

A science fiction drama television series on CBS that focuses on a secret government project investigating the first contact with an extraterrestrial species. (September 16, 2005–February 1, 2006; CBS, Sky1).

### The Twilight Zone (Original)

*"There is a fifth dimension beyond that which is known to man. It is a dimension as vast as space and as timeless as infinity. It is the middle ground between light and shadow, between science and superstition, and it lies between the pit of man's fears and the summit of his knowledge. This is the dimension of imagination. It is an area which we call The Twilight Zone."* You betcha! *TZ* presented an eclectic mix of tales that were part sci-fi, part fantasy, part morality tale, part O. Henry, part early Steven Spielberg(!), and included stories about aliens, other worlds, other dimensions, and occult and paranormal phenomena. Truly a unique television creation, it is quite possible that the brilliance of the original *Twilight Zone* series has yet to be duplicated—even though Rod Serling helmed a *TZ*-like series called *Night Gallery* from 1970–1973. *Night Gallery* was good, but nothing came close to the original six seasons of *The Twilight Zone*. (October 2, 1959–September 26, 1965; CBS).

### UFO

This British sci-fi series was set in the future—1980!—and told of the efforts of the super, mega, *ultra* top secret organization SHADO (The Supreme Headquarters Alien Defense Organization) to defend the Earth against the newly-diskovered and ongoing, deadly threat from UFOs. (SHADO was so secretive because the general public did not know they were under attack by aliens.) Like *Star Trek, UFO* had its share of pastel-colored aliens and futuristic hardware, and it aired in syndication in the United States for one season. (1972 – 26 episodes; Syndicated).

### V

This weekly series had its genesis as two very popular mini-series, and it told the stories of the Visitors ("V" for "Visitors")—friendly aliens who came to Earth wanting our minerals, in exchange for which they would solve all our problems with their advanced technology. We Earthlings bent over, the V made themselves at home, and before long our benefactors and new neighbors became our dictators and were ruling the world. A resistance organization evolved, and each week an attempt was made to wrest control of our home planet away from the Visitors. (October 26, 1984–July 5, 1985; NBC; November 3, 2009–March 15, 2011; ABC).

### V (Reboot)

*V* is a reboot of the science fiction television miniseries dealing with the arrival of a highly-advanced alien species claiming to come to Earth in peace. As is usually the case, this turns out to be the opposite of the truth. (November 3, 2009–March 15, 2011; ABC).

### Voyage to the Bottom of the Sea

The *Seaview* was a glass-nosed atomic submarine that patrolled the deepest depths of Earth's oceans, defending the planet from alien life forms that wanted to take over the world. This show was based on Irwin Allen's 1961 movie of the same name and was one of the sixties' most popular series. (September 14, 1964–September 15, 1968; ABC).

### War of the Worlds

According to this series, the 1938 invasion of Earth by aliens (Welles's radio broadcast) had been a reconnaissance mission, and the 1953 war (the Gene Barry movie) had been with aliens from the planet Mortax. This series picked up thirty-five years after the first war when the aliens hidden in storage in Nevada escape and go on the attack. In the second season of the show, Earth was being attacked by beings from the planet Morthrai who, it turns out, had been using the aliens from the first season as soldiers. The Morthrai executed the Mortaxians in the first episode of season two for failing in their mission. (1988–1990; Syndicated).

### Warehouse 13

US Secret Service Agents Myka Bering and Pete Lattimer are assigned to Warehouse 13 in South Dakota, a mysterious compound intended to hold supernatural artifacts. (July 7, 2009–May 19, 2014; SyFy).

### The X-Files

In the introduction to the *Playboy* Interview with David "Fox Mulder" Duchovny in the magazine's December 1998 issue, *The X-Files* is described as "a cross between *Twin Peaks* and *The Twilight Zone*—a TV series for paranoids and zealots who are sure the government covers up what it knows about the UFOs and aliens among us." Series star Duchovny was a tad more laconic: "We do a cop show with paranormal phenomena," he told interviewer Lawrence Grobel. However you describe it, there is no denying that *The X-Files* is a phenomenon unlike almost anything else produced for television. We learned early on in the series that Duchovny's character, Mulder, became obsessed with aliens and the paranormal after he saw his sister abducted by aliens when she was eight. Mulder's partner, Dana Scully (Gillian Anderson), was the skeptic. Together they investigated the FBI's

"X-Files"—cases that defied conventional explanations. A true cult hit, a 1998 *X-Files* movie—*The X-Files: Fight the Future*—was a box-office smash. The show has generated countless books, websites, and conventions and was created (and written) by Chris Carter. (September 10, 1993–; Fox).

### The X-Files (Revival)

While officially the tenth season of the venerable sci-fi show, the return of *The X-Files* more than fourteen years after the conclusion of its ninth season, and seven years after the last film in the franchise most certainly qualifies as an important sci-fi show. Especially for those viewers too young to remember just how big *The X-Files* was in its day. (January 24, 2016–present; Fox).

# ACKNOWLEDGMENTS

We'd like to thank Mike Lewis, John White, Joe Craig, Tony Lyons, Bill Wolfsthal, and all the fine folks at Skyhorse Publishing. Also, we'd like to express our gratitude to Brian Lesmes, researcher and writer, who came through with some superb research on a very tight clock. Also, our profuse thanks to Rachel Montgomery, who probably now knows more about Roswell than she ever thought she would want to know.